A DARK ENERGY THEORY CORRELATED WITH LABORATORY SIMULATIONS AND ASTRONOMICAL OBSERVATIONS

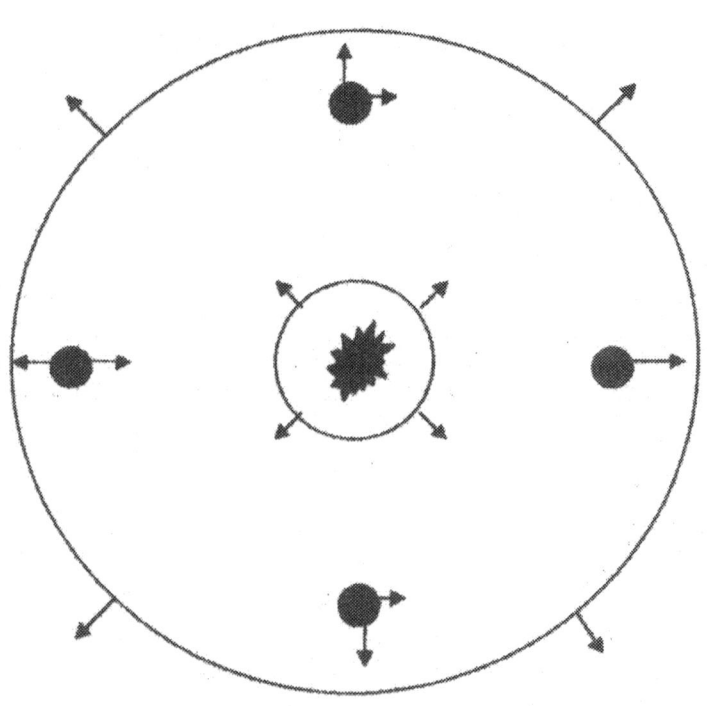

A DARK ENERGY THEORY CORRELATED WITH LABORATORY SIMULATIONS AND ASTRONOMICAL OBSERVATIONS

August A.Cenkner Jr.
B.A., B.S., M.S., Ph.D.
University at Buffalo
Buffalo, New York
USA

First published by AuthorHouse 08/23/05

ISBN: 1-4208-3447-9 (sc)

Library of Congress Control Number: 2005902703

Printed in the United States of America
Bloomington, Indiana

This book is printed on acid-free paper.

authorHOUSE™

1663 LIBERTY DRIVE, SUITE 200
BLOOMINGTON, INDIANA 47403
(800) 839-8640
WWW.AUTHORHOUSE.COM

To my wife Judy

She suggested that I write this book. Her boundless energy and enthusiasm, after all these years, never ceases to amaze and inspire me.

CONTENTS

Preface xiii

1 Introduction 1

2 Presentation

   2.1 Unexplained Phenomena 3

      2.1.1 Star Acceleration -- In Outer Reaches Of The Universe 4

      2.1.2 Stars Are Not Distributed Uniformly In The Universe 5

      2.1.3 Groups Of Revolving Stars Or Dark Bodies 6

   2.2 Dark Energy Explained 7

   2.3 Traveling Shock Wave Created In Shock Tube Compared To Shock

      Wave From Exploding Star 8

   2.4 What Happens When A Traveling Shock Wave Impacts A Star 9

   2.5 Laboratory Simulation To Study Impact Of Shock Wave On Star 10

      2.5.1 Process To Be Simulated --- Traveling Shock Wave Impacts On Star..11

      2.5.2 Simulation Of Shock Impact On Star Using Plasma Wind Tunnel 12

      2.5.3.1 Plasma Wind Tunnel 13

      2.5.3.2 Schematic Of Plasma Wind Tunnel 14

      2.5.4.1 Plasma Wind Tunnel Test Section 15

      2.5.4.2 Schematic Of Plasma Wind Tunnel Test Section 16

      2.5.5.1 Unique Optical System For Measuring 3-D Temperature Distributions

      In Plasma 17

      2.5.5.2 Calibration Of Film And Alignment Of Mirror Systems 18

      2.5.6 Side Image Of Plasma With Film Calibration Data 19

      2.5.7.1 Simultaneous Plasma Images From 12 Different Angles -- Run 427....20

      2.5.7.2 Simultaneous Plasma Images From 12 Different Angles -- Run 431... 21

      2.5.7.3 Deflection Of Plasma Gas By Shock Wave 22

      2.5.8 Constant Temperature Lines In Plasma Symmetric Cross Section 23

      2.5.9.1 Effect Of Shock Strength And Electric Field Strength On Plasma

      Structure -- Side Views 24

CONTENTS

2.5.9.2 Effect Of Shock Strength And Electric Field Strength On Plasma
Structure -- Isotherms 25

2.5.10 Temperature Comparison Between Main Sequence Stars And
Laboratory Plasma 26

2.5.11 Observations On Plasma Wind Tunnel Testing 27

2.6 High Pressure Shock Wave Creating Pressure Differential On Star Or Dark
Body 28

2.7 Newton's Second Law Of Motion 29

2.8 Application Of Newton's Law -- Velocity Increase For Given Impulse 30

2.9 Measured Radial Velocities Of 3,500 Stars 31

2.10 Energy Released From Various Violent Processes 32

2.11 Acceleration Of Incompressible Stars And Dark Bodies (p=150,000 "Little
Boy" atomic bombs) 33

2.12 Effect Of Uniform Spherical Star Compression On Acceleration (p=150,000
"Little Boy" atomic bombs) 34

2.13 Observations 35

2.14 Spherical Shock Wave Moves Stars And Dark Bodies In Various Directions 36

2.15 Dark Energy Zone Explained 37

2.16 Creation Of Superclusters, Voids And Walls 38

2.17 Creation Of Revolving Groups Of Stars, Dark Bodies, And Nebula Vortex 39

2.18 Correlation Of Dark Energy Theory Predictions With Astronomical
Observations 40

2.19 Conclusions 41

3 Dark Energy Explained 43

4 Shock Wave Flow Past Star Or Dark Body 45

5 Acceleration Predictions Using Newton's Law 46

5.1 Impulse Predictions 46

5.2 Incompressible Bodies 47

5.3 Compressible Bodies 48

CONTENTS

6 Other Unexplained Phenomena 49

    6.1 Galactic Superclusters, Walls, and Voids 49

    6.2 Unusual Star Motion 50

    6.3 Revolving Star And Dark Body Groups 51

    6.4 Nebula Vortex 51

7 Conclusions 51

8 Acknowledgements 53

9 References 54

TABLES

1 Summary Of Measured Radial Star Velocities 56

2 Energy Released And Shock Strength Of Various Violent Processes 56

FIGURES

1 High Pressure Shock Wave Creating Pressure Differential On Star Or Dark Body 57

2 Impulse Required For Velocity Increase (0.5-1.25 Ms)....................................... 57

3 Impulse Required For Velocity Increase (10-25Ms) 58

4 Acceleration Of Incompressible Star And Dark Bodies (p=1 "Little Boy" atomic
    bomb) 58

5 Acceleration Of Incompressible Star And Dark Bodies (p=150,000 "Little Boy" atomic
    bombs).. 59

6 Effect Of Star Compression On Star Acceleration (p=1 "Little Boy" atomic bomb) 59

7 Effect Of Star Compression On Star Acceleration (p=150,000 "Little Boy" atomic
    bombs) 60

8 Spherical Shock Wave Moves Stars In Different Directions 60

APPENDIX

History Of Dark Energy Theory 61

PREFACE

I was getting bored, really bored. As a consultant, there are times when there are relatively long periods between contracts; this was one of those periods.

After a while, I began thinking that maybe this was in fact a great opportunity for me. I could actually do anything I wanted to do -- but what.

My technical background is comprised of interdisciplinary training and experience, including the areas of gas-dynamics/physics/astronomy/computers. In addition, I have been involved in theoretical and experimental research, or development, for about thirty-three years. After all these years, I actually had the opportunity to select the type of research that I wanted to perform; in the past, someone else always made that decision. This was great!

I started searching for projects that were of interest to me, that were compatible with my background, and that I could handle with the resources that were available to me. My limited resources included: myself, a computer, the internet, local university libraries, and amateur astronomy groups -- but no outside funding.

Eventually, I decided there were two areas where I might be able to contribute: random processes – as applied to turbulent gas flows and random vibrations – and astronomy. I started pursuing both.

I joined the Buffalo Astronomical Society, a group of very enthusiastic amateur and super-amateur astronomers. They have a sophisticated observatory and they were encouraging members to use it. Some of their members were already involved in variable star research and in astrophotography. My original plan was to use their observatory to get involved in the search for new planets, using the transit approach. I would eventually employ a spectroscopic technique to search for absorption lines during planet transit, to see if I could identify the presence of a planet atmosphere.

But then I started reading about dark energy. What a "cool" name – dark energy. During all my years of teaching and R&D, I had heard about a lot of different types of energy but never dark energy. Becoming fascinated with the concept, I conducted an internet search and uncovered three papers that caught my attention:

Chaikin, A., Dark Energy: Astronomers Still Clueless About Mystery Force Pushing Galaxies Away, 2002, Space & Science

Villard, R. & Lloyd,, R., Astrophysics Challenged by Dark Energy Finding, 2001, Space.com

Weinberg, S., Importance of Discovering the Nature of Dark Energy, Department of Physics, 2001, University of Texas at Austin, http://supernova. lbl.gov/~evlinder/weinberg.pdf

With my theoretical and experimental R&D background, how could I resist. I was hooked! I immediately dropped all other work and concentrated on dark energy. Drawing on my interdisciplinary background, I gradually evolved the dark energy theory presented herein.

Initially, I was inspired by original experimental research that I had conducted 35 years ago for my M.S. thesis, which was for an unrelated application. I revisited the experimental results when it became apparent that this experiment could actually be reinterpreted as a small scale laboratory simulation of dark energy -- as it starts to accelerate a main sequence star. The rest of the theory evolved from this point by applying my gas dynamics/physics/astronomy background.

During the course of my study, a review was made of an excellent descriptive cosmology book that is entitled "Cosmology Revealed -- Living Inside The Cosmic Egg"; it was written by Professor Anthony Fairall. I just couldn't forget one plot that showed how galaxies are non-uniformly distributed. Especially fascinating was the observation that Voids are spherical-like in shape, with galactic Clusters and galactic Walls forming around these Voids. Once I decided what dark energy is, it dawned on me -- dark energy could also be responsible for creating these structures, in addition to causing stars to accelerate. It was obvious that theoretical predictions were consistent with these observations.

This book would be of specific interest to anyone interested in astronomy, cosmology, astrophysics, or gas dynamics. However, it was written so that it would also be of interest to serious amateur astronomers and especially to those with interdisciplinary skills. Hopefully they will also become fascinated with astronomy research and become involved in the many areas that are waiting to challenge them.

August A. Cenkner Jr.

A DARK ENERGY THEORY CORRELATED WITH LABORATORY SIMULATIONS AND ASTRONOMICAL OBSERVATIONS

August A. Cenkner Jr.

A Dark Energy Theory Correlated With Laboratory Simulations And Astronomical Observations

1. INTRODUCTION

An explanation is introduced to place a classical face on the elusive dark-energy, the currently unknown repulsive force that is held responsible for the acceleration of galaxies in the outer regions of the universe. An application of Newton's Law of Motion shows that traveling shock waves can account for the observed galactic accelerations. The theory shows that dark-energy may also have contributed to the creation of other unexplained phenomena like revolving star/dark body groups, Superclusters, Voids, Walls, and nebula vortex. In addition, it is shown that shock waves can contribute to drag on stars/dark bodies and to the creation of wild-stars. The predictions are reinforced by laboratory simulations of dark-energy that demonstrate what will happen when a traveling shock wave impacts a star. A qualitative correlation with observations shows that the dark energy explanation is consistent with the limited available observational data.

In 1998, by studying the emission spectra from galaxies in the outer reaches of the universe, two independent teams of astronomers concluded that these distant galaxies have actually accelerated; they are moving at higher, rather than the anticipated lower, velocities (Villard & Lloyd 2001; Chaikin 2002; Weinberg 2001; Krauss 2004). Classical thought is that gravitational attraction, between celestial objects, should actually slow them down instead of speeding them up. They have attributed this unexpected behavior to some mysterious repulsive force that has been labeled dark-energy. The nature of this dark-energy is presently unknown. In addition, there are other currently unexplained phenomena like revolving star/dark body groups, Superclusters, Voids, Walls, and nebula vortex. These phenomena are also discussed and shown to be related to dark-energy.

1

An unconventional format is used for this book. Section 2 was inspired by a well-received dark energy presentation that was made at Buffalo State College in Buffalo, NY. The presentation began with a brief review of some currently unexplained astronomical phenomena, as shown in Section 2.1. In Section 2.2, dark energy is explained; it is argued that it is actually high energy traveling shock waves. An analogy is then drawn between the behavior of a conventional shock tube and an astronomical shock emitter, like an exploding star or galaxy.

As noted in Section 2.4, a key question is "What happens when a traveling shock wave hits a star (or dark body)?" To answer this question, laboratory testing on a plasma wind tunnel – that was performed for unrelated reasons – is revisited and reinterpreted in an astronomical context. Using spectroscopic and photographic diagnostics, it is shown what can be expected to happen when a traveling shock wave impacts a star. It is noted that the actual response will obviously depend upon the size of the star and the strength of the shock wave.

Using the resulting qualitative description of a star/shock wave interaction, Newton's Law of Motion is employed to show how shocks can produce significant star and dark body acceleration. It is then shown how spherical shock waves can account for the unexplained phenomena that was discussed earlier. A correlation is made between qualitative predictions of the shock wave theory and reported astronomical observations. These correlations are summarized in Section 2.18.

Sections 3-7 discusses, in more detail, the information outlined in the presentation. Finally, the history of the attempted dissemination of the dark energy theory is reviewed in the Appendix.

2

2. Presentation

2.1 Unexplained Phenomena

- Universe Acceleration
- Galactic Superclusters
- Voids
- Walls
- Revolving Star and Dark Body Groups
- Wild Stars

2.1.1 Unexplained Phenomena -- Star Acceleration In Outer Reaches Of The Universe

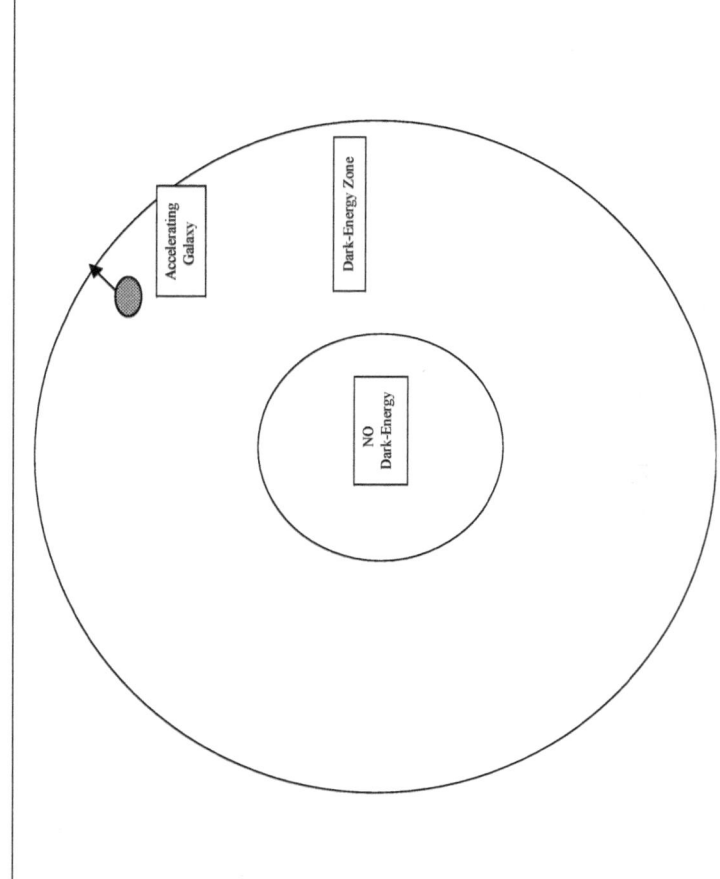

2.1.2 Unexplained Phenomena -- Stars Are Not Distributed Uniformly In Universe*

Non-uniform distribution of:
- Galaxy Clusters
- Voids (spherical shapes)
- Walls

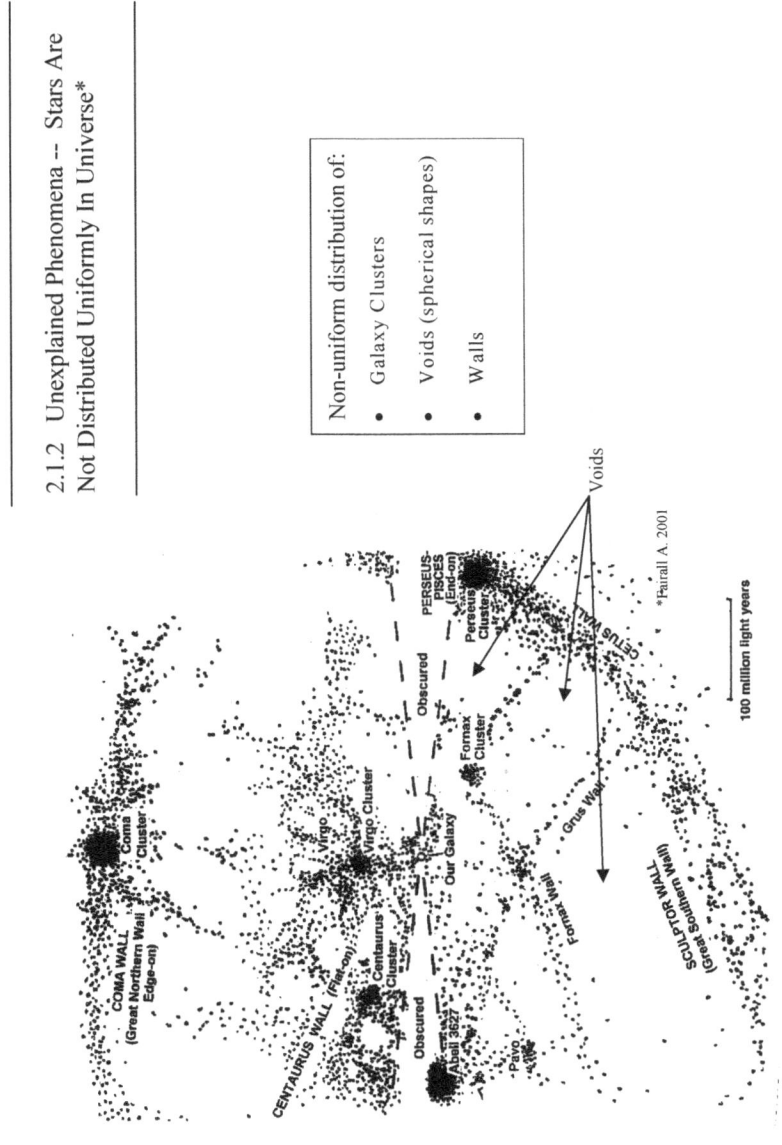

*Fairall A. 2001

100 million light years

2.1.3   Unexplained Phenomena – Groups of Revolving Stars or Dark Bodies

Why did stars, dark bodies and gas/particle clouds start revolving?

C. M.

2.2 Dark Energy Explained

- Dark Energy

  - Energy contained in traveling shock waves

    - Acceleration created when shock wave impacts star or dark body

    - Acceleration continues until shock wave passes

- Processes that create shock-emitters

  - Galactic explosions

  - Exploding or imploding stars

  - Supernova

  - Pulsating stars

  - Black holes, as matter approaches the event horizon

  - Etc

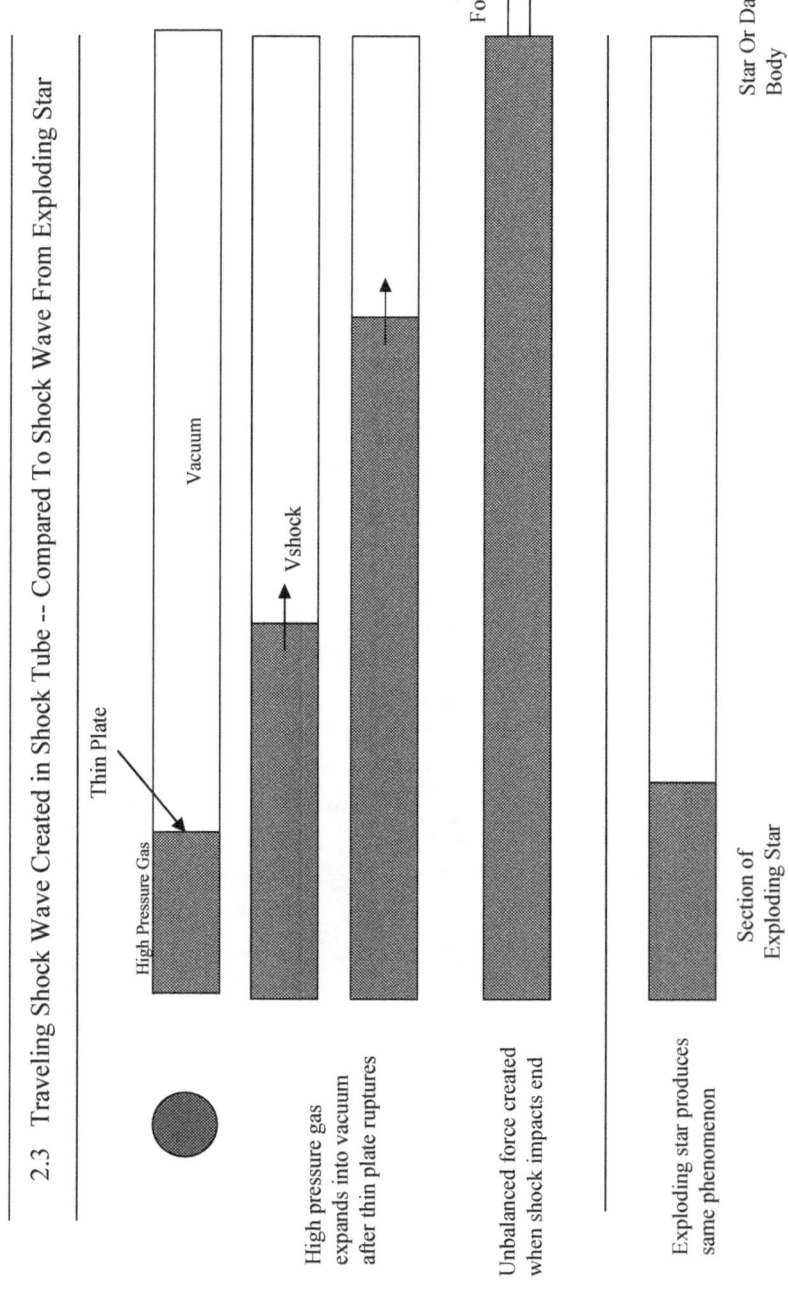

2.3   Traveling Shock Wave Created in Shock Tube -- Compared To Shock Wave From Exploding Star

Thin Plate

Vacuum

High Pressure Gas

High pressure gas
expands into vacuum
after thin plate ruptures

Vshock

Unbalanced force created
when shock impacts end

Force

Exploding star produces
same phenomenon

Section of
Exploding Star

Star Or Dark
Body

2.4 What Happens When A Traveling Shock Wave Impacts A Star

- Passes through star?

- Absorbed by star?

- Ignored by star?

- Completely envelopes star, producing no net force on star?

- Passes around star, producing a net force on the star?

2.5 Laboratory Simulation To Study Impact Of Shock Wave On Star

- High temperature plasma wind tunnel employed
  - Plasma created by electric arc
- Simulate gravitational field by electric field
- Spectroscopically measure 3-D temperature distribution inside plasma
  - Determine what happens when shock impacts on plasma

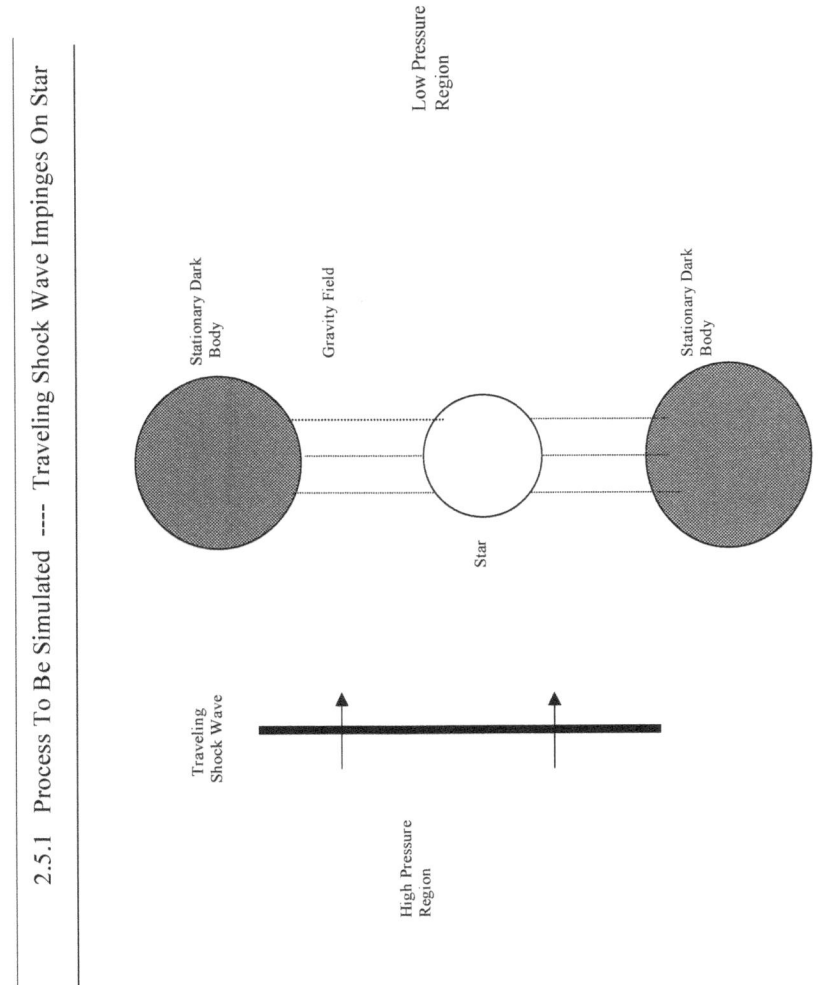

2.5.1 Process To Be Simulated --- Traveling Shock Wave Impinges On Star

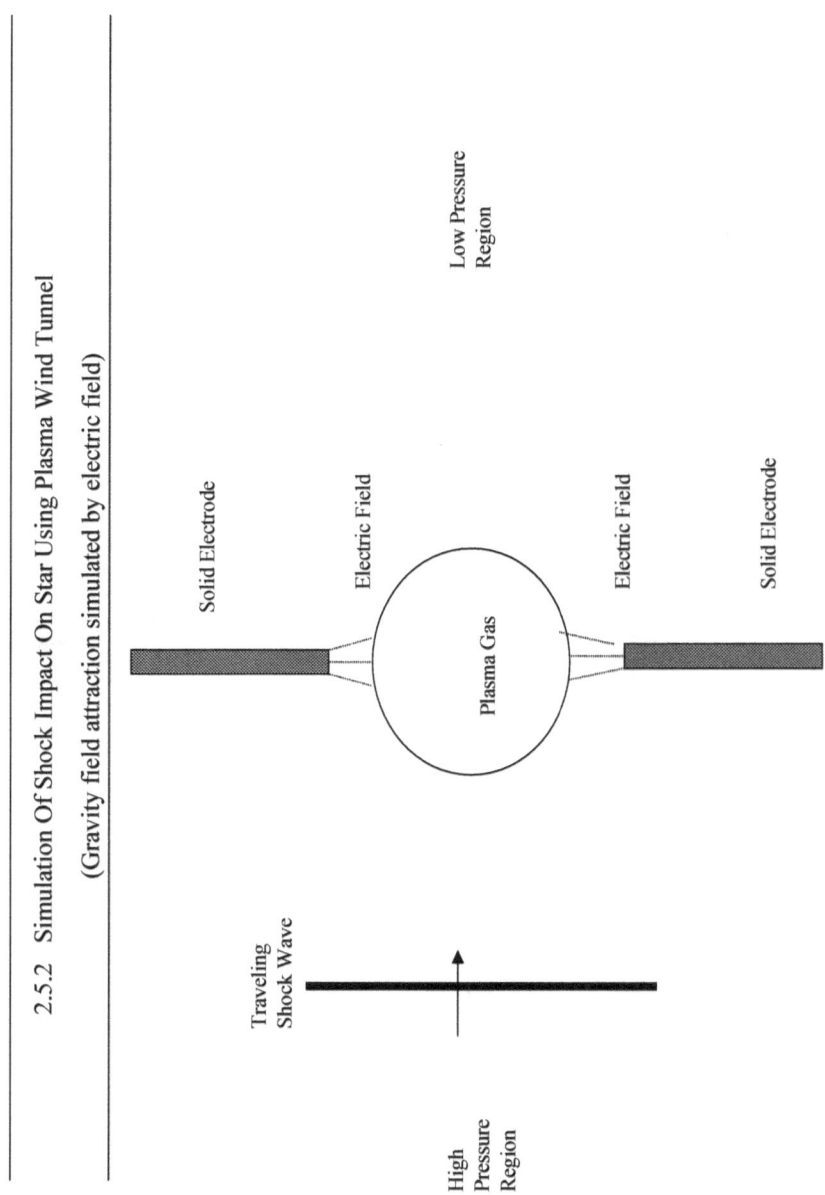

2.5.2   Simulation Of Shock Impact On Star Using Plasma Wind Tunnel

(Gravity field attraction simulated by electric field)

2.5.3.1   Plasma Wind Tunnel

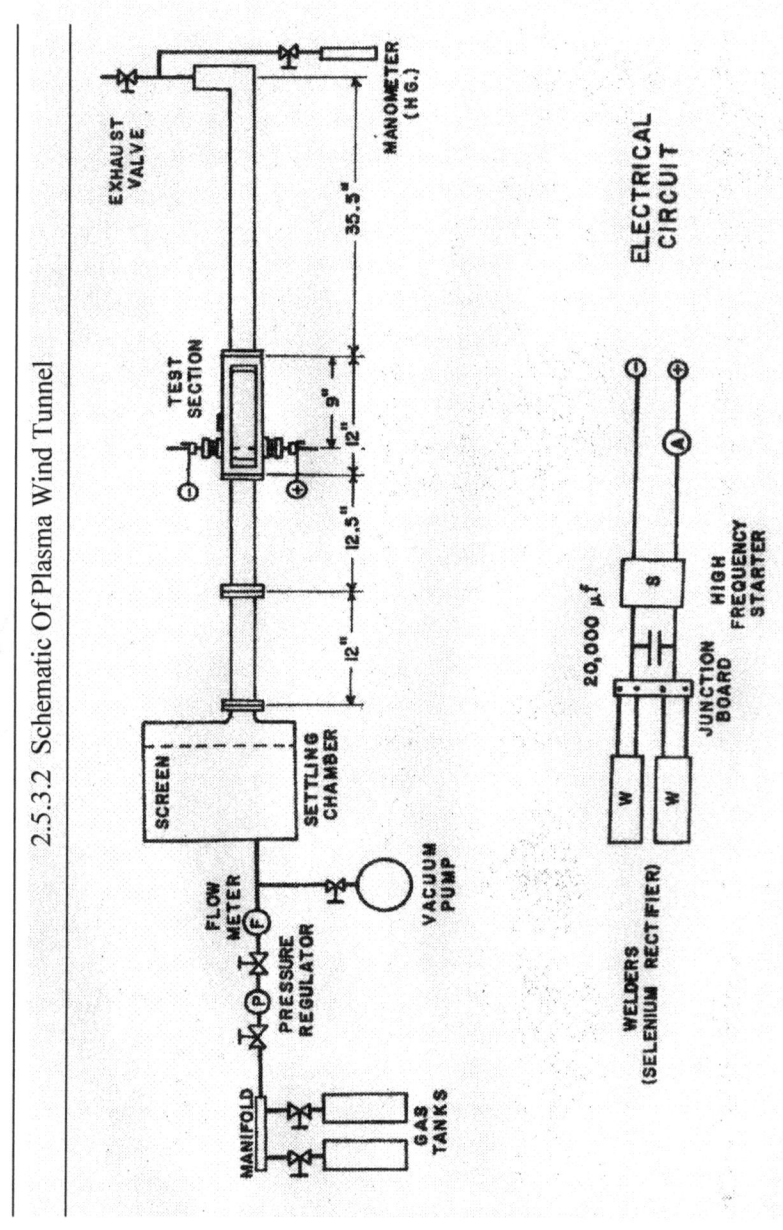

2.5.3.2 Schematic Of Plasma Wind Tunnel

2.5.4.1 Plasma Wind Tunnel Test Section

2.5.4.2 Schematic Of Plasma Wind Tunnel Test Section

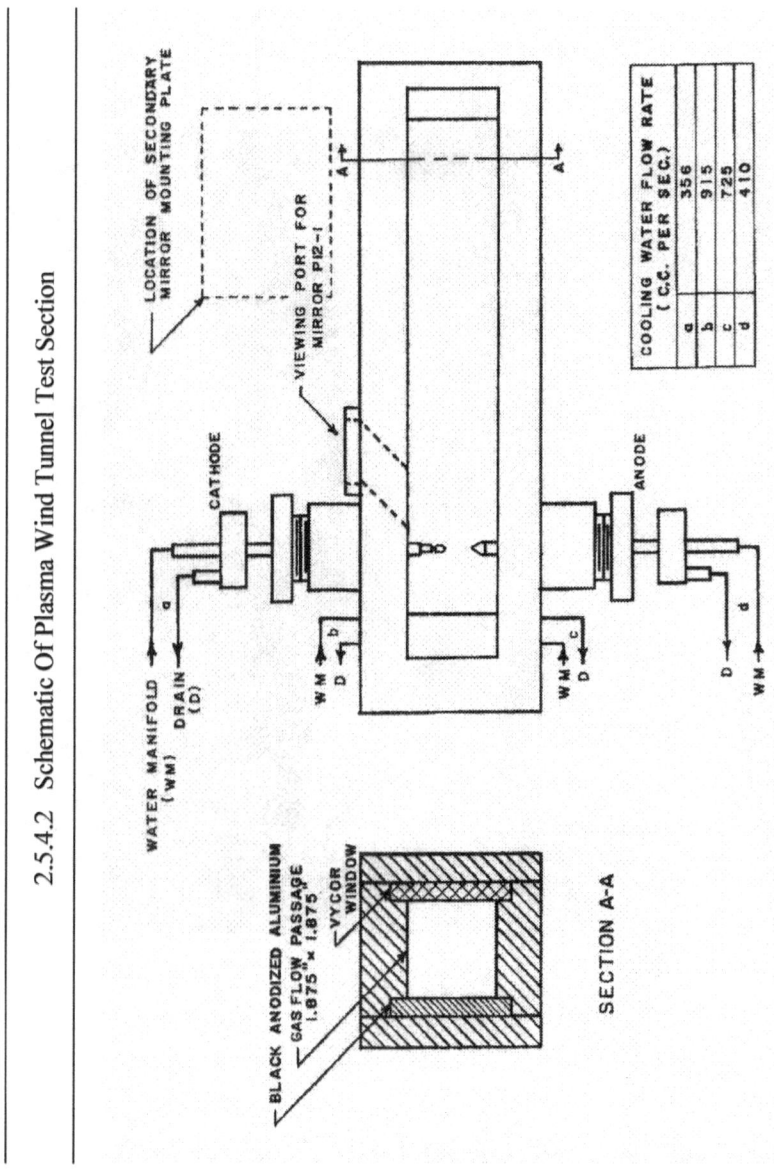

2.5.5.1 Unique Optical System For Measuring 3-D Temperature Distributions In Plasma

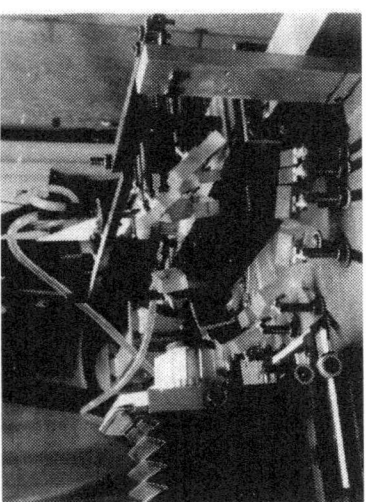
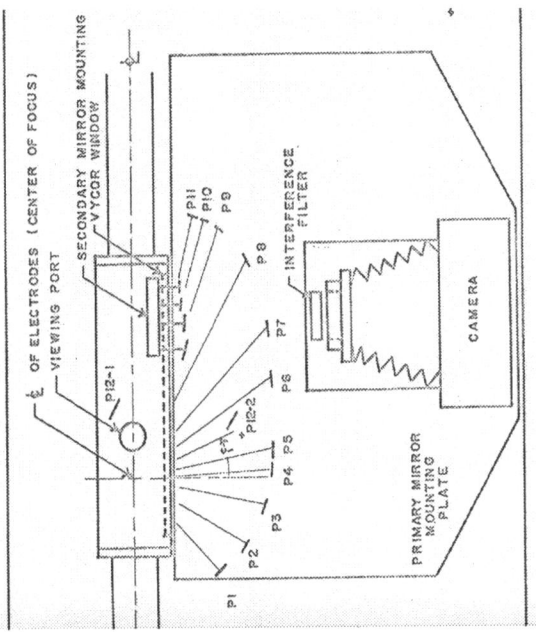

2.5.5.2   Calibration Of Film And Alignment Of Mirror Systems

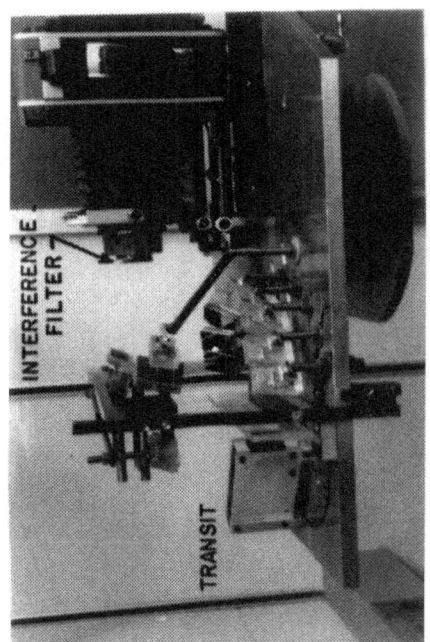

Measuring Viewing Angle Of Each Mirror System

Placing Calibration Images On Film

2.5.6   Side Image Of Plasma With Film Calibration Data

Electrode

High temperature plasma -- deflected by shock wave

Electrode -- with groove for spatial orientation

"Black body" images -- used to calibrate photographic film
using densitometer scans

(Taken through same optical system that was used for
plasma image)

August A. Cenkner Jr.

2.5.7.1 Simultaneous Plasma Images From 12 Different Angles -- Run 427

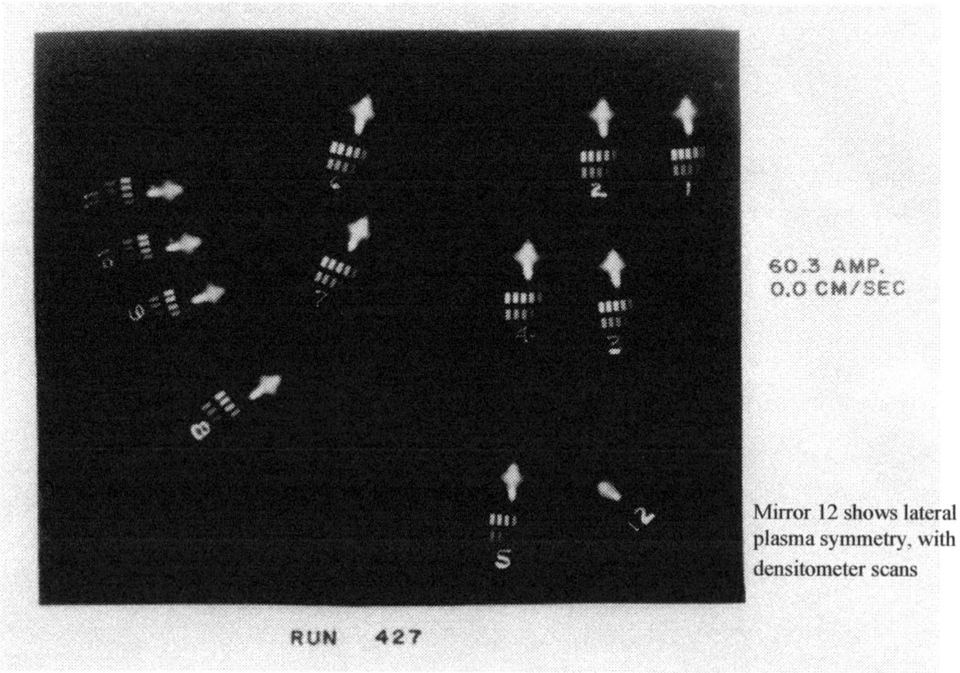

60.3 AMP.
0.0 CM/SEC

RUN 427

Mirror 12 shows lateral plasma symmetry, with densitometer scans

20

2.5.7.2 Simultaneous Plasma Images From 12 Different Angles -- Run 431

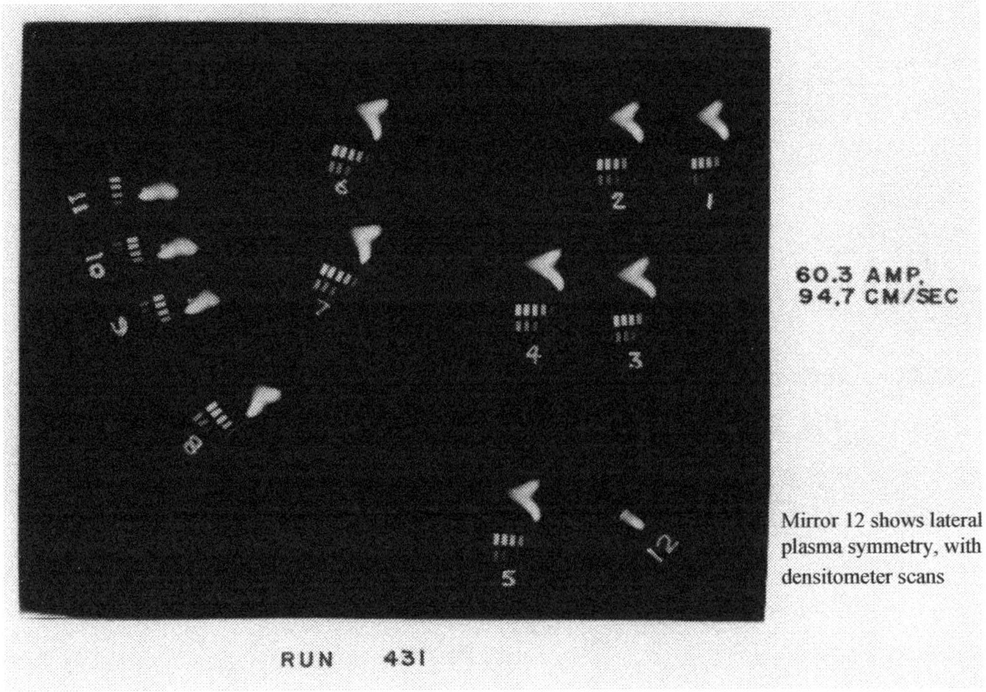

60.3 AMP.
94.7 CM/SEC

RUN 431

Mirror 12 shows lateral
plasma symmetry, with
densitometer scans

2.5.7.3 Deflection Of Plasma Gas By Shock Wave

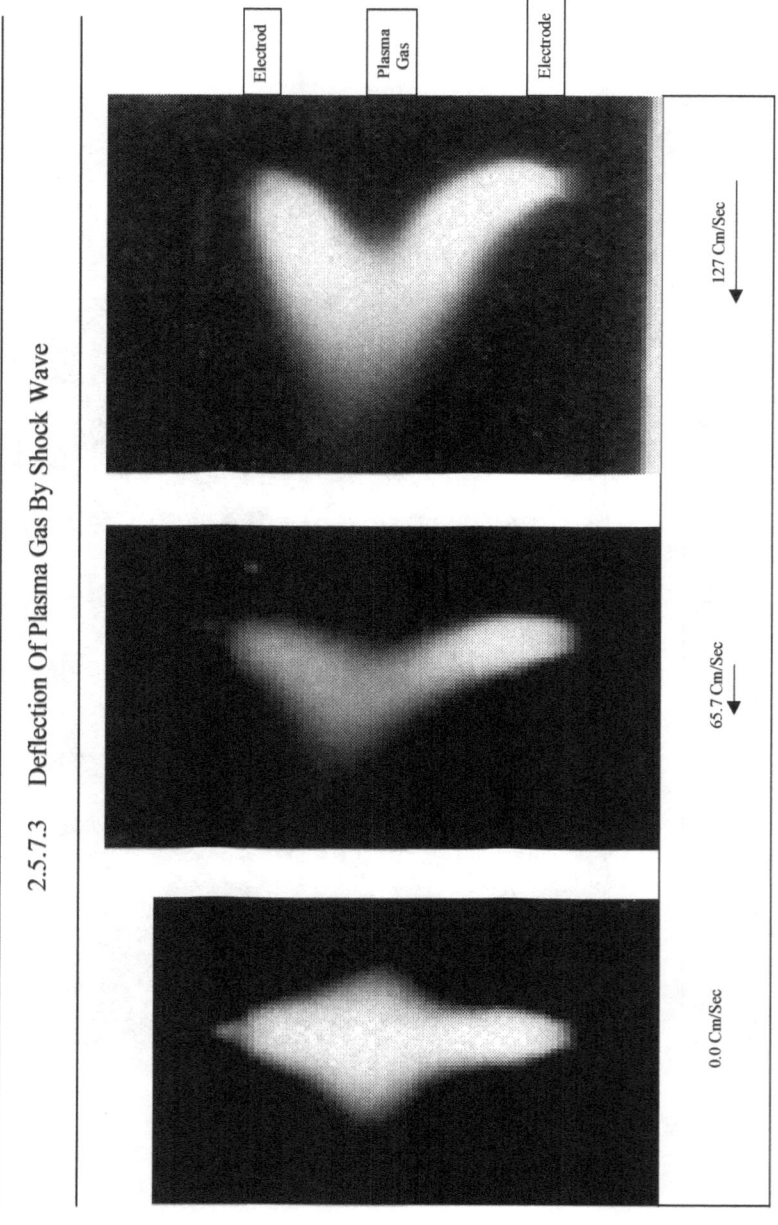

2.5.8 Constant Temperature Lines In Plasma Symmetric Cross Section

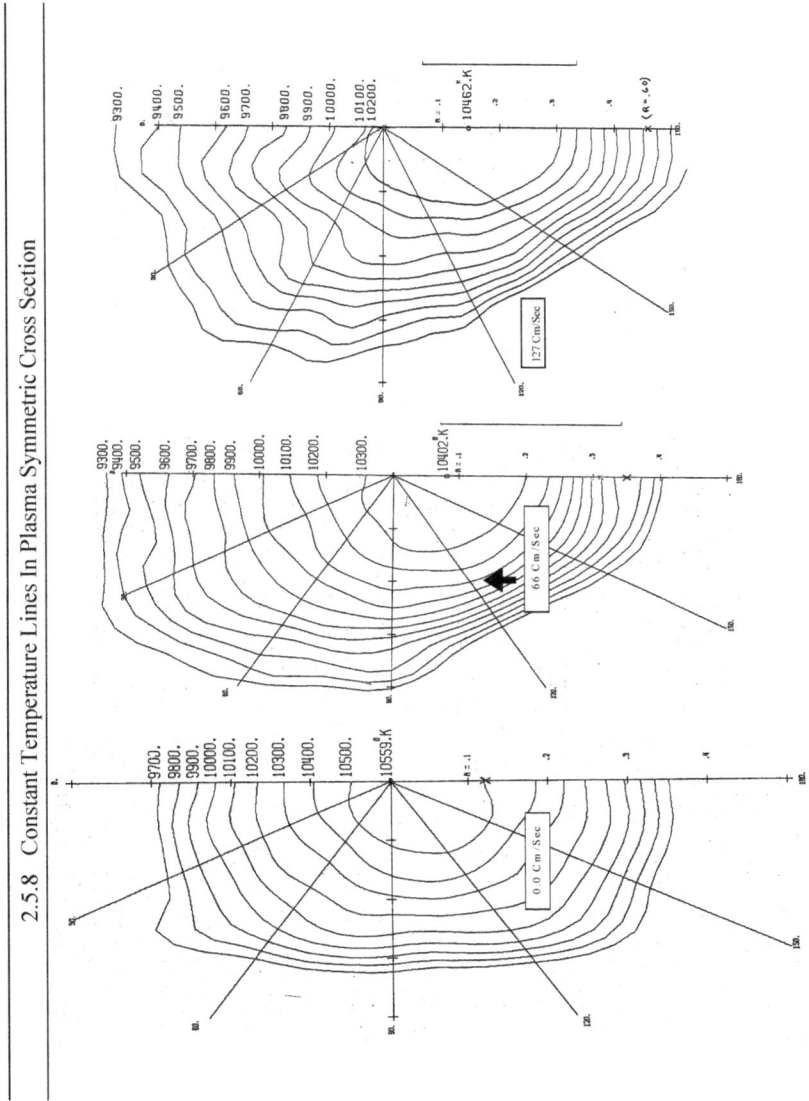

23

2.5.9.1   Effect Of Shock Strength And Electric Field Strength On Plasma Structure -- Side Views

2.5.9.2 Effect Of Shock Strength And Electric Field Strength On Plasma Structure -- Isotherms

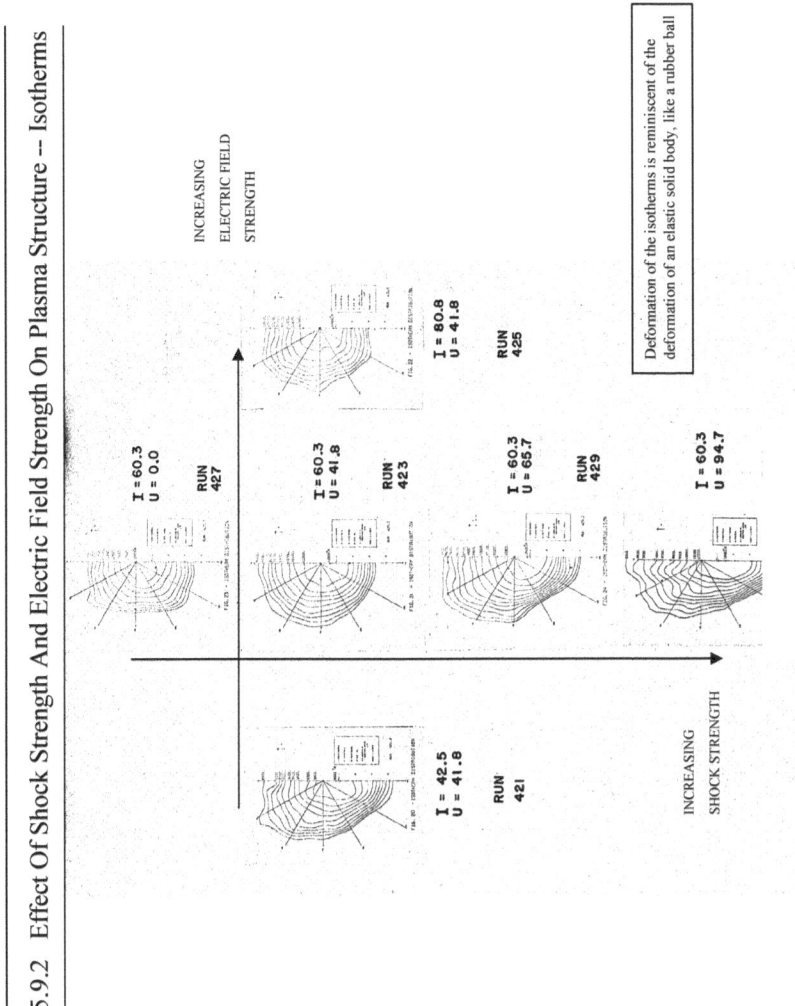

25

2.5.10 Temperature Comparison Between Main Sequence Stars And Laboratory Plasma**

MAIN SEQUENCE STARS

Spectral type	Temp (k)	Mass*	Radius*
O5	40,000	40	17.8
B0	28,000	18	7.4
A0	9,900	3.2	2.5
A5	8,500	2.1	1.7
G0	6,000	1.1	1
K5	4,100	0.7	0.7

*Multiples of solar mass and radius

MEASURED LABORATORY PLASMA TEMPERATURE

Run No.	Current (amps)	Mean Gas Velocity (cm/sec)	Max Temp (k)	Min. measured isotherm Temp (k)
427	60.3	0	10,559	9,700
423	60.3	41.8	10,333	9,400
429	60.3	65.7	10,402	9,300
431	60.3	94.7	10,462	9,300
433	60.3	127.0	10,593	9,700
421	42.5	41.8	9,940	8,900
423	60.3	41.8	10,333	9,400
425	80.8	41.8	10,800	10,000

** The plasma surface temperature is essentially the same as the surface temperature of some main sequence stars

2.5.11 Observations -- Plasma Wind Tunnel Testing

- Shock Wave Does Not Pass Through Plasma Gas
  - Plasma Behaves Like a Solid Deformable Body
- Plasma Gas Is Deflected Downstream By Shock Wave But Remains Symmetric Laterally
  - There is a unbalanced force acting on the plasma
  - This implies there is a low pressure wake downstream of plasma
  - Plasma goes out when shock flow is high enough
  - This simulates star breaking free of gravitational attraction

2.6 High Pressure Shock Wave Creating Pressure Differential On Star Or Dark Body

2.7 Newtons' Law Of Motion

$$F_{net} = M \, dV/dt \qquad \text{eq. 1}$$

$$\text{Impulse} = F_{net} \times Dt = M \times DV = M \times (V2-V1) \qquad \text{eq. 2}$$

$$F_{net} \times Dt = P \times (Pi \times R \times R) \times Dt = M \times DV \qquad \text{eq. 3}$$

2.8　Application Of Newton's Law -- Velocity Increase For Given Impulse

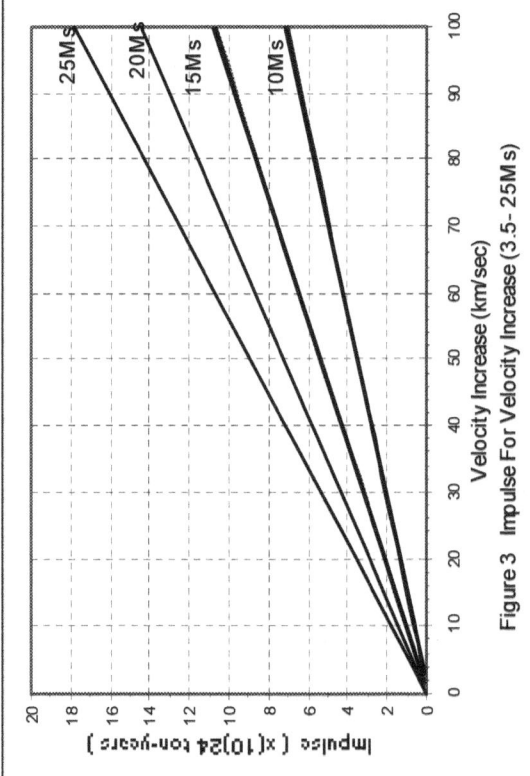

Figure 3　Impulse For Velocity Increase (3.5- 25Ms)

(1) Various strength shock waves, with different duration, produce same velocity increase if have same impulse.

(2) Multiple shock waves produce same velocity increase, if total impulse is the same.

30

2.9 Measured Radial Velocities Of 3,500 Stars*

	Velocity Range 11000 velocities of 3500 stars (km sec^{-1})		
	Low	High	Isolated max
Southern Hemisphere (R.A. : 0 − 12^{h})	+0.2	+236	+555
	−1.0	−142	none
Northern Hemisphere (R.A. : 0 − 12^{h})	+0.4	+298	+346
	−0.4	−326	−414

* statistical analysis of data from Abt 1970

2.10   Energy Released From Various Violent Processes*

Source of Explosion	Energy Released (joules)	Equivalent TNT (tons)	Shock Pressure (tons m^{-2})
1 ton TNT	4.20 E+09	1	-------
"Little Boy" Atomic Bomb	5.25 E+13	1.25 E+04	35
Supernova Type II	1.00 E+46	2.38 E+36	?

*A supernova II explosion contains considerably more energy than 150,000 " Little Boy" atomic bombs

2.11   Acceleration Of Incompressible Stars And Dark Bodies (p=150,000 "Little Boy" atomic bombs)

(1) After 20 days, a star 3.2 times the mass of the sun (Ms) would have a velocity increase of about 130 km/sec.

(2) With a Supernova Type II explosion, it would take a lot less than 20 days, because it has so much more energy than 150,000 "Little Boy" atomic bombs (Section 2.10)

2.12   Effect Of Uniform Spherical Star Compression On Acceleration (p = 150,000 "Little Boy" atomic bombs)

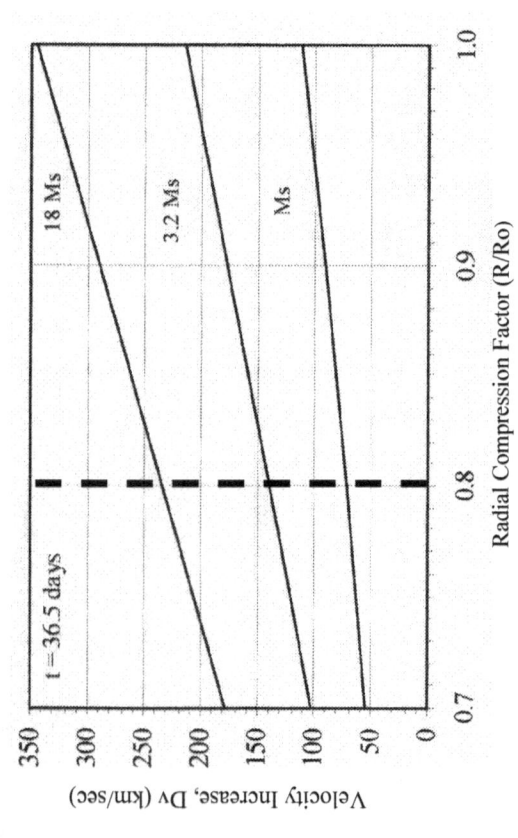

(1) For a star 3.2 times as massive as the sun, uniform compression reduction of its radius to 80%, reduces the velocity increase from 210 to 140 km/sec.

2.13 Observations

- Traveling Shock Waves Can Produce Significant Acceleration Of Stars And Dark Bodies

  - No Attempt Was Made To Predict How Much Acceleration Would Occur

    - Unknowns

      - Temporal Shock Strength $(P(t))$

      - Shape and Size Of Star After Shock Impact

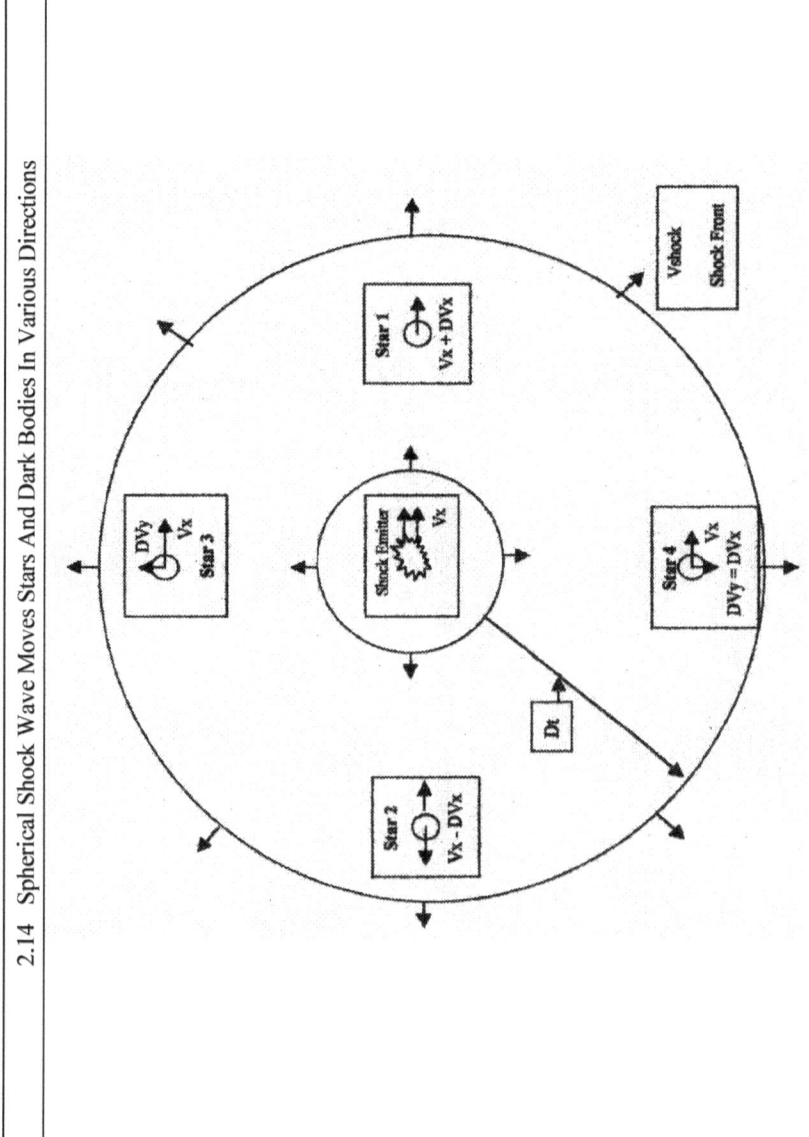

2.14 Spherical Shock Wave Moves Stars And Dark Bodies In Various Directions

2.15   Dark Energy Zone Explained

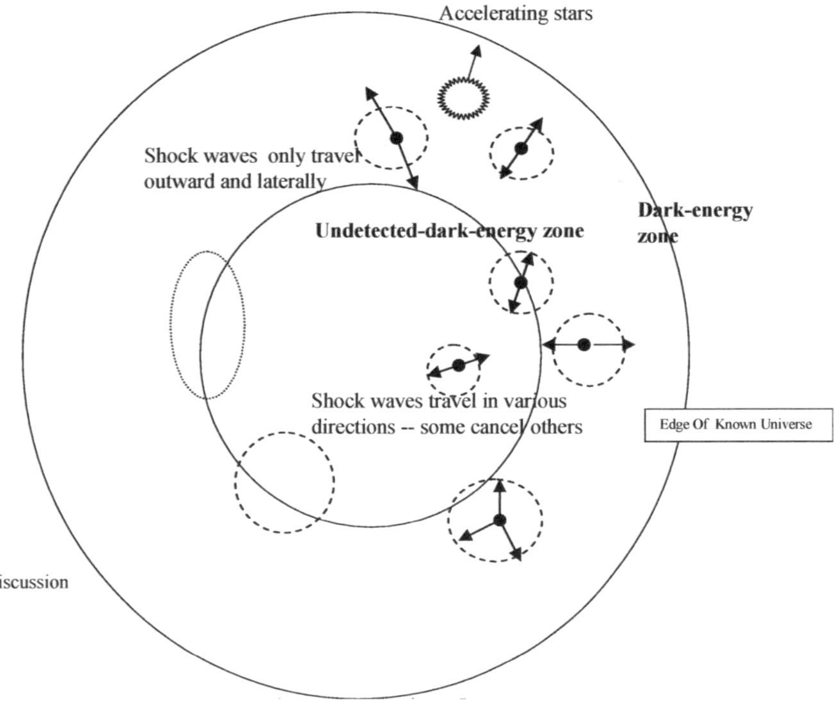

Accelerating stars

Dark-energy zone

Edge Of Known Universe

Undetected-dark-energy zone

Shock waves only travel outward and laterally

Shock waves travel in various directions -- some cancel others

See Section 3 for discussion

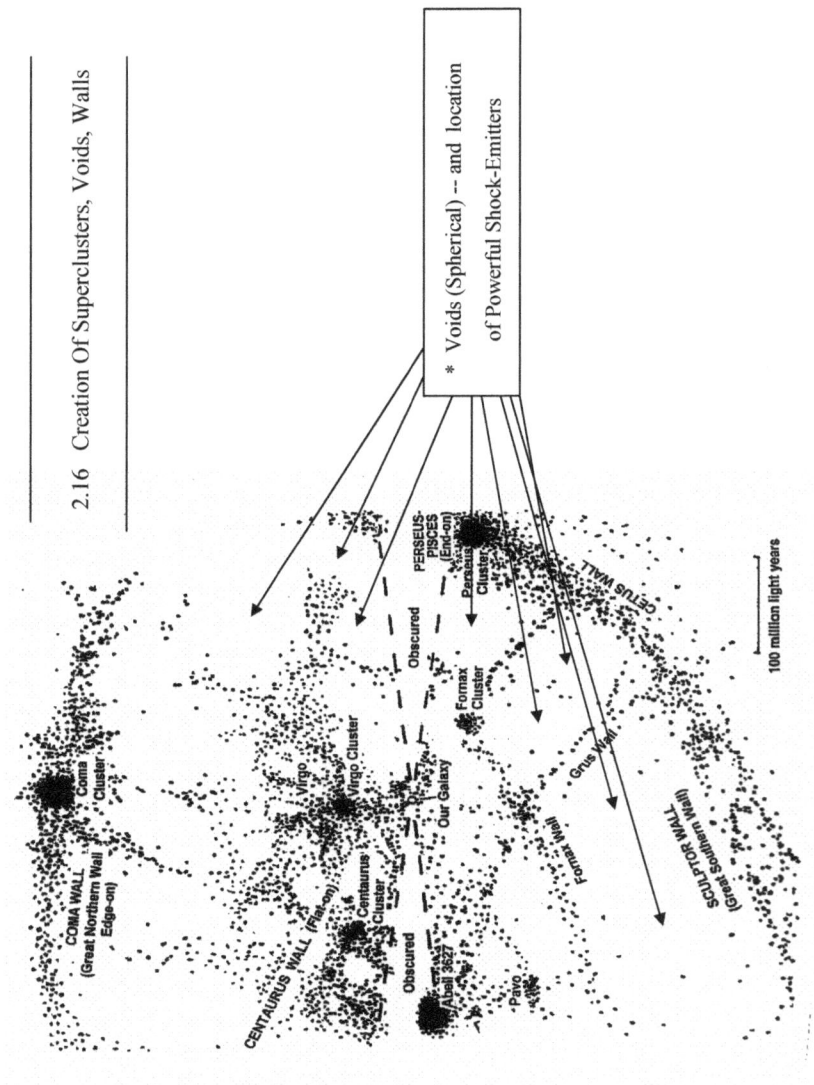

2.16 Creation Of Superclusters, Voids, Walls

* Voids (Spherical) -- and location of Powerful Shock-Emitters

2.17 Creation Of Revolving Groups Of Stars, Dark Bodies, And Nebula Vortex

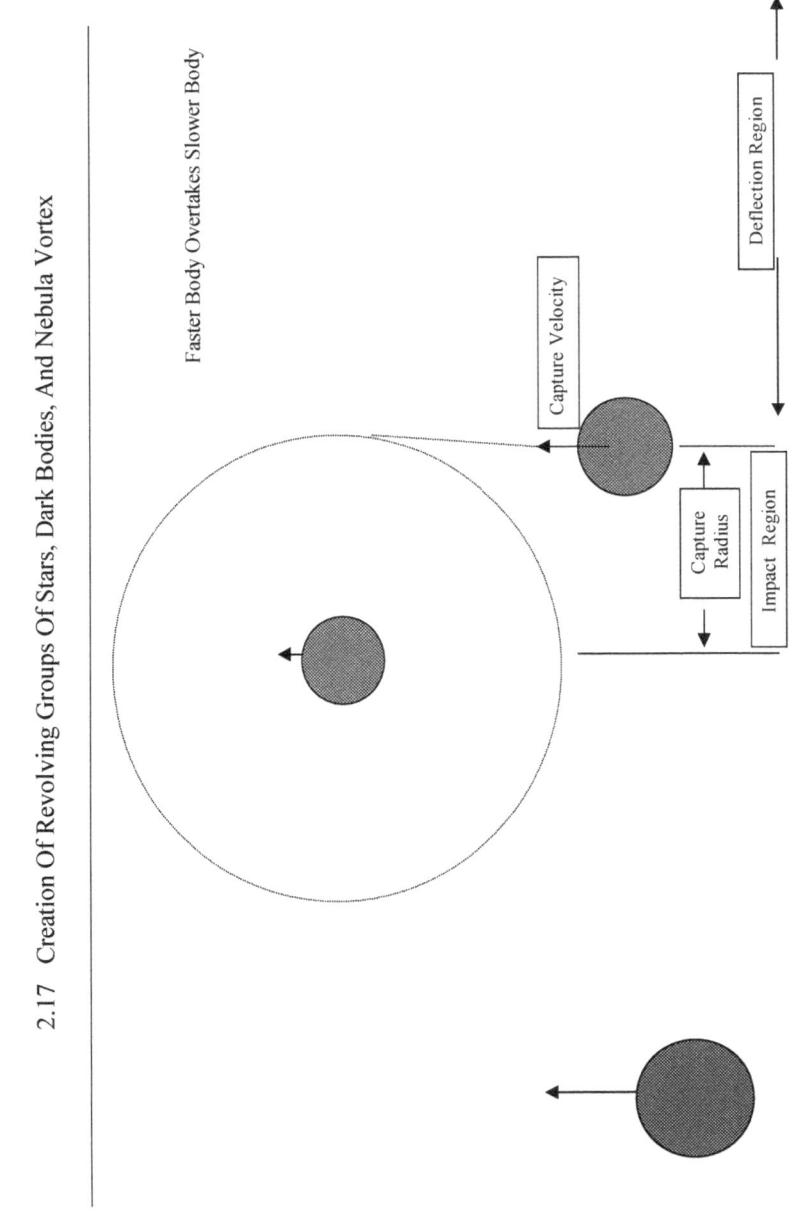

Faster Body Overtakes Slower Body

Capture Velocity

Deflection Region

Capture Radius

Impact Region

2.18 Correlation Of Dark Energy Theory Predictions With Astronomical Observations

Phenomena	Qualitative Dark Energy Theory Predictions For Stars And Dark Bodies	Astronomical Observations
Star acceleration in outer reaches of the universe	Predicted – shock waves can only travel outward and laterally in this region	Observed – attributed to Dark Energy
Star acceleration in inner regions of the universe	Predicted – if there are not multiple shock waves to cancel acceleration from first shock	*Observed –not attributed to DarkEnergy
Voids	Predicted – spherical shock waves move bodies, in all directions, to create Voids	Observed
Galactic Clusters	Predicted – when multiple shock waves move in various directions and cancel out their effects on the bodies, when merging	Observed
Galactic Walls	Predicted – when multiple shock waves move in various directions and cancel out there effects on the bodies, when merging	Observed
Groups revolving around common center of mass – bodies can revolve in opposite directions	Predicted – under the right conditions, the shock results in various bodies being captured by other bodies	Observed
Particle/gas clouds revolving about center of mass	Predicted – under the right conditions, the shock results in various particles being captured by other particles	Observed
Deceleration of bodies	Predicted – if shock impact occurs on front of body, the body will slow down	Observed -- but attributed to gravitational forced or drag
Wild Stars – individual or groups of bodies translating opposite to universe expansion direction	Predicted – if shock wave is strong enough to overcome initial momentum of body	Unknown
Lateral motion of bodies, relative to direction of expansion of the universe	Predicted – shock waves can move bodies laterally	Observed – requires more study

* In 1572 the Tycho star – part of a binary star system -- went supernovae. Recently (Reddy 2005), it was found that the companion star is moving through the supernovae remnants at a velocity that is about three times that of nearby stars. According to Sections 4 and 5 of this book, this acceleration can be caused by the supernovae shock wave, i.e. dark energy.

2.19 Conclusions

- For The First Time, A Classical Theory Is Presented To Describe How Unexplained Phenomena Occured

  - Traveling Shock Waves Can Produce The Unexplained Phenomena

- Traveling Shock Wave Theory Predictions Are Consistent With Observational Data

- Predicting How Much Star Acceleration Will Occur, And For How Long, Requires Extensive Computer Modeling

  - Sophisticated 3-D Computer Model Of Violent Processes That Create Shock-Emitters

    - Shock Strength Vs Time

- Sophisticated 3-D Computer Model Of Shock(s) Impinging On Star And Dark Bodies

    - Must Include Compressibility Of Star

    - Should Include Shock Reflections

Conclusions -- continued

- Cross-flow plasma wind tunnels can be used to obtain quantitative information on interaction of shock waves and stars

  - Various strength shock waves
  - Various size stars
  - Electric field strength simulates gravity field strength
  - Use nondimensional parameters

- Photographic technique upgraded to system of CCD cameras

3. DARK-ENERGY EXPLAINED

A dark-energy explanation is presented herein, that predicts the observed acceleration of astronomical bodies in the outer reaches of the universe.

The basic theory states that multiple hypervelocity traveling gas-pressure/radiation-pressure/stellar-wind shock waves are producing numerous force impulses when they impinge on an astronomical body. If the body is in front of the shock-emitter, these impulses would cause it to speed up. If it is behind the shock-emitter, it would slow down. The shock waves would be moving at a velocity that is higher than the emitter velocity, which would enable the shock to overtake bodies moving in front of the emitter. Bodies traveling behind the emitter would slow down because of retarding shock impulses.

As will be demonstrated in Sections 4 and 5, this acceleration behavior can be predicted by applying Newton's Law of Motion. The unbalanced pressure on a star, that occurs upon shock impact, will produce a force impulse [force x time] that will increase the linear momentum [mass x velocity] of the astronomical body and hence cause it to accelerate to a higher velocity.

These traveling shock waves will also affect the velocity and path direction of nebula, (and possibly stimulate star formation) as these waves pass through the nebula. Nebula that are in front of the shock emitter would be accelerated while nebula that are behind it would be decelerated. Since there could be more than one shock-emitter in a galaxy or star group, these multiple shock waves would interact. The shocks could be reflected, attenuated or enhanced and could produce multiple or enhanced shock impulses.

The so-called dark energy is actually the energy contained in these high-energy hypervelocity traveling shock waves. The thermal kinetic and internal energy of the star is converted into the kinetic and internal energy of the traveling shock wave during some type of violent process. When the shock impinges on a star, or other body, it performs work on the body. This work increases the kinetic and internal energy of the body and reduces the shock energy.

Possible sources of these shock waves include galactic explosions, exploding stars, pulsating stars, imploding stars, colliding stars, black holes as matter approaches the event horizon, stellar wind, gamma ray bursta, etc. This shock wave energy may permeate the entire universe over time, since these shock-emitters may appear, eventually, wherever the proper type stars exist.

It has been reported that the core of the universe doesn't presently exhibit the presence of dark energy. Rather, it occurs in an expanding spherical shell around the core. This seems to suggest that shock emitters in the core region may be the major emitters, with galactic explosions, supernova, cataclysmic variables, and gamma ray bursts, being the prime candidates as the most powerful of the relevant shock-emitters, but not necessarily the most dominant. This dark energy then travels from the core toward the outer regions, as shown in Section 2.15.

There are a number of scenarios that could explain the existence of a dark energy zone, for example:

Scenario 1: (Some of) the stars in the "dark energy zone" are actually 2nd, 3rd, or higher generation stars that evolved from the remnants of shock forming events, sometime in the past. Even though this region of the universe is the oldest, the stars are young relative to the age (and size) of more mature stars that can transform into shock-emitters. Alternatively, stars in the "no detected dark energy zone" are mature enough to become shock emitters. However, these shock-emitters in the inner zone may not have been currently identified as the source of the dark energy that travels to the outer zone.

Scenario 2: Current observation data is incomplete. Acceleration in the "dark energy zone" is really localized, due to localized outward bound shock waves.

Some stars in this region are being influenced by inward bound shock waves, but are currently undetected because they are localized.

The theory suggests that, eventually, all shock-emitters will be consumed. Gravity will again become the dominant force and drag may more significantly influence cosmic motion. The expansion of the universe may slow down, if gravity is significant enough and the bodies are close enough.

Of course, some of the remnants of the shock-emitters may eventually coalesce into a new generation of stars. Some of these stars may then eventually transform into a new generation of shock-emitters. This would enlarge the dark-energy zone and prolong expansion.

4. SHOCK WAVE FLOW PAST STAR OR DARK BODY

To see how dark energy causes stars and solid dark bodies to accelerate, it is necessary to first consider what happens when a traveling shock wave collides with a star; see Sections 2.4 & 2.5.

Consider a high pressure (H.P.) shock wave moving past a star, or dark body, that is situated in a low pressure (L.P.) environment. The shock wave is moving at a velocity Vshock and the star is initially moving at a velocity Vx.

When the shock impinges on the star, the star is compressed somewhat, the flow separates and passes around the star, a stagnation point is created on the forward side of the star and gas dynamic and thermal boundary layers are formed at the surface of the star, as in Figure 1. Due to the compression of the shock on the forward side, various waves are probably created. (Laboratory experiments have determined that a high temperature plasma, which has a gaseous cross-flow impinging on it, will deform like a 3-D deformable body and the flow will separate and pass around it; as reported by (Benenson, Cenkner 1970a, 1970b, 1971). In the direction of flow, the cross-section eventually becomes elliptical. See Section 2.5 in the Presentation, for details.

When the shock passes the star, a low-pressure region, the wake, still exists behind the star. This is because the high-pressure shock medium cannot expand fast enough to fill in the wake. This creates a pressure, and hence force, differential on the body -- which causes it to accelerate.

5. ACCELERATION PREDICTIONS USING NEWTON'S LAW

5.1 IMPULSE PREDICTIONS:

To determine the acceleration of the star or dark body, after shock wave impact, Newton's Law of Motion is applied:

$$Fnet = M \, dV/dt \qquad\qquad \text{eq. 1}$$

If the shock pressure, and therefore Fnet, is essentially constant over the time impact interval Dt, this equation can be integrated to give:

$$\text{Impulse} = Fnet \times Dt = M \times DV = M \times (V2\text{-}V1) \qquad \text{eq. 2}$$

Using equation 2, the impulse required to produce various velocity changes, as computed for the main-sequence stars of the H-R diagram (Seeds 2004), is shown in Figures 2 & 3. The velocity increase range was selected by statistically analyzing the measured total radial velocities of 3500 stars; see (Abt 1970). These identified velocity ranges are summarized in Table 1. Figures 2 & 3 therefore give shock impulses that would produce significant accelerations of main sequence stars, based on these reported measured radial velocities.

Figures 2 & 3 highlight several interesting phenomena. First, different multiple shock waves, that have various shock strengths and impinge for different time intervals, will produce the same increase in velocity if they impart the same total

46

impulse magnitude. Next, either a single shock wave or multiple shock waves will produce the same velocity increase, if they exert the same total impulse. Finally, It would require different magnitude impulses to increase the velocity, of different size bodies, by the same amount.

5.2 INCOMPRESSIBLE BODIES:

The preceding impulses can be computed without making any assumptions about the composition of the star. However, several assumptions have to be made in order to be able to determine the shock strength required to produce the corresponding velocity increases:

(1)   It will first be assumed that the body is essentially incompressible; this would apply to degenerate stars, solid dark bodies, and weak shock waves impacting on main sequence stars but would be more of an approximation to higher strength impacts on main sequence stars. In the next section this assumption will be relaxed and the case of a compressible star or gaseous dark body will be addressed.

(2)   Both the compressible and incompressible body assumptions will also assume that the shock pressure P is constant over the time interval Dt.

With the assumption that the body remains spherically rigid during shock impact, with a radius R, Newton's Law becomes:

$$Fnet \times Dt = P \times (Pi \times R \times R) \times Dt = M \times DV \qquad eq. 3$$

To determine the required shock duration, to create a given velocity increase, it is necessary to know the temporal shock strength, P (t). Jones 2004, reports that supernova may persist for more than 1,000 years. Unfortunately, there is no reliable information available on temporal shock strength. Nevertheless, in order to bound the problem and to gain some insight into what traveling shock waves are capable of doing, it is useful to use the known shock strength of a violent high-energy explosion. No Author 2004, reports a ground shock strength of 35 tons per square meter from a 12.5 kiloton TNT atomic bomb, called "Little Boy". While this ground strength would be stronger than that obtained from the same bomb if it was exploded in space, it is a useful reference. It is illuminating to compare to total energy released in a Type II supernova explosion (Jones 2004) and that released by a "Little Boy" atomic explosion, which is summarized in Table 2. This shows that the "Little Boy" impulse prediction will provide a very low bound on a body acceleration caused by a supernova. It would also give a lower bound for less powerful shock-emitters. For example, according to Figure 5, after 20 days a star or dark body that is 3.2 times the mass of the sun (Ms) would have a velocity increase of about 130 km/sec. However, with a Supernova Type II explosion, it would take a lot less than 20 days, because it has so much more energy than 150,000 "Little Boy" atomic bombs (Table 2).

Figure 4 gives the shock duration that would produce significant accelerations for a shock having the strength of a single "Little Boy" ground shock wave, as obtained from equation 3. The corresponding data for 150,000 "Little Boy" shock waves is presented in Figure 5.

Figures 4 & 5 reveal that a given strength shock wave will produce different velocity increases for different size stars, regardless of the duration of the impact. For example Figure 5 predicts a velocity increase, after 20 days impact of a shock with the strength of 150,00 "Little Boy" atomic bombs, of 45 km/sec for a 0.5 Ms body and a higher velocity increase of 185 km/sec for an 18 Ms body.

5.3 COMPRESSIBLE BODIES:

When a main sequence star is compressed by a higher strength shock wave impinging upon it, the projected- pressure-area is reduced. The net applied force and impulse are also reduced, while the total mass remains unchanged. To explore the effect of compressibility, it was assumed that the star is compressed uniformly and it retains a spherical shape on the leading edge side. Figures 6 & 7 reveal what happens when the star radius is reduced uniformly by compression. When the radius is reduced to 70% of its uncompressed value, the star-velocity-increase is reduced by 49%. While this shows there still is a significant acceleration prediction, it also demonstrates that compressibility effects have to be taken into account if the goal is to predict as accurately as possible the amount of acceleration and not just that there is an acceleration.

6. OTHER UNEXPLAINED PHENOMENA

The application of Newton's Law of Motion in Section 4, to the study of the interaction of traveling shock waves and various astronomical bodies, can be used to gain some insight into other hereto-unexplained phenomenon.

6.1 GALACTIC CLUSTERS, WALLS AND VOIDS:

It has been determined that the universe doesn't have a uniform 3-D distribution of stars and galaxies (Fairall 2001). Rather, the stars can gather together to form groups or clusters and the galaxies can group together to form Galactic Clusters. Some galaxies also group in such a way as to form long stringy ensembles called Walls. Large regions called Voids, that are essentially free of stars, separate these clusters. These voids are spherical-like in shape and they come in various sizes, from extremely large to very small as shown in Section 2.1.2.

At present, there is no theory to explain how these Clusters, Walls, and Voids were formed.

The theoretical predictions of Section 4 indicate that dark energy (i.e. high energy transient traveling shock waves) could also have contributed to the creation of these star Clusters, as well as the Voids and Walls. To see how dark energy could have helped create these large-scale structures, consider Figure 8. Assume a powerful shock-emitter is embedded in a uniform field of stars (or galaxies). Now consider how the resulting traveling shock wave will interact with the four stars shown in Figure 8. The shock-emitter and the stars are all moving with an initial velocity Vx. A violent phenomenon occurs and the central star is converted into a shock-emitter. This creates a spherical high-speed high energy traveling shock wave that exists for time interval Dt. The unbalanced high-pressure shock impacts on the stars, causing them to accelerate, as was demonstrated in Section 5. The magnitude of the velocity increase, DVx, is determined by the shock strength and the duration of the shock impulse, Dt. Star 1 increases in velocity by DVx , star 2 decreases by DVx , while stars 3 and 4 are pushed sideways at a velocity DVy = DVx. Since this happens to all the stars inside the spherical shock, a sphere-like Void is formed. Once the trailing edge of the shock passes, gravity and drag forces take over and begin to control the motion of the stars -- unless another shock wave passes through the region.

Different size Voids are created by multiple shocks that have different strengths, durations, points of origin, and originate from different size stars or star groups. The appearance of shock-emitters in nearby Voids can also affect a given void size because the interaction of the shocks will limit the spreading of each one. This is illustrated in Section 2.16.

With powerful enough shocks, or multiple shocks, whole galaxies may be moved in this way, creating the galactic Clusters or Superclusters. The Walls may be formed when powerful shock waves, traveling in opposite directions, impinge upon the same galaxies from opposite directions; each shock wave cancels the effect of the other.

6.2   UNUSUAL STAR MOTION:

Assume that star 2 in Figure 8 is a "free floating" star or solid dark body, that is not revolving around any other body. If the shock-emitter is strong enough -- so that DVx is larger than the initial body velocity Vx -- star 4 can actually reverse direction. It can travel in a direction that is opposite to the direction of expansion of the universe, becoming what could be called a "wild-star".

6.3   REVOLVING STAR AND DARK BODY GROUPS:

Figures 4 and 5 demonstrate that a given strength shock wave will accelerate different size stars and dark bodies, in a field of bodies, to different velocities. The faster bodies will eventually overtake the slower bodies. Some bodies will collide and some will just have their trajectories changed. If the faster body is within the capture range of the slower body, and it is moving with the capture velocity, it will begin orbiting around the slower body. Both bodies will then orbit around their common center of mass (Section 2.17). A second, or more, higher speed bodies could then possibly be captured also. It's conceivable that eventually a group of revolving stars, a group of revolving dark bodies, or even a group of dark bodies revolving around a star, would evolve. Since faster bodies could be passing on either side of the center of mass, it would be possible for different bodies to revolve in different directions around the center of mass.

6.4   NEBULA VORTEX:

Consider a nebula comprised of gas and various size particles. As the shock wave passes through the nebula the gas is compressed and accelerated. The different size suspended particles are also accelerated to different velocities. The faster particles will again overtake the slower particles and some of them will begin orbiting the slower particles, if they possess the proper capture parameters (Section 2.17). More and more particles will be captured and revolve about their center of mass, forming a particle vortex. Due to the frictional interaction between the revolving particles and the gas, the particles will drag the gas with them. This will establish a gas/particle vortex mixture, which will continue to grow.

7. CONCLUSIONS

Newton's Law of Motion was used to predict the acceleration of stars and other astronomical bodies by, the impact of high energy traveling shock waves.

Correlation with limited observational data, that is summarized in Section 2.18, has led to the conclusion that these shock waves are the elusive dark-energy that has been reported in the literature. It was also concluded that these shock waves could be responsible for other unexplained phenomena like revolving star and solid and gaseous dark body systems, Clusters, Superclusters, Voids, Walls, nebula vortex, and wild-stars.

No attempt was made in this study to quantify the acceleration; much additional challenging work would be required by a number of teams. It would entail the use of sophisticated computer models to simulate various violent processes, which would create shock-emitters, in order to evaluate their shock strength and duration and to identify the dominant source(s). It would also require sophisticated 3-D computer models of traveling shocks impinging upon compressible and incompressible astronomical bodies. For correlation, accurate measurement of star and dark body accelerations would have to be made, in the dark-energy-zone.

8. ACKNOWLEDGEMENTS

The experimental research was conducted at the Plasma Physics Laboratory, at the University of Buffalo, Buffalo, NY, on Contract AF33(615)-1797 for the Aerospace Research Laboratories, Air Force Systems Command, United States Air Force. It was also partially supported by the National Science Foundation Grant GK-1174, the Research Foundation of the University at Buffalo, and the Computing Center at the University of Buffalo. The Computing Center was partially supported by NIH Grant FR-OO126 and NSF Grant GP-73l8. Dr. David Benenson, a professor at the University at Buffalo, was the principal investigator on the experimental research that was done in the Plasma Physics Laboratory in 1969.

9. REFERENCES

Abt, A.A. , Catalog of Individual Radial Velocities, 0h-12h, Measured by Astronomers of the Mount Wilson Observatory, 1970,   Kitt Peak National Observatory

Benenson, D. M., &   Cenkner Jr, A. A. , Effects of Velocity & Current on Temperature Distributions Within Cross-Flow Electric Arc, 1970a, Transactions of ASME Journal of Heat Transfer, Vol 92

Benenson, D. M. &   Cenkner Jr,, A. A. , Three-Dimensional Temperature Distribution Within a Steady-State Cross-Flow Arc, 1970b, Aerospace Research Laboratories, Office of Aerospace Research, United States Air Force, ARL 70-0135

Cenkner Jr, A. A. , Steady-State Characteristics Of A Cross-Flow Electric Arc, 1969, Master of Science Thesis, University at Buffalo, Buffalo, NY

Chaikin, A., Dark Energy: Astronomers Still Clueless About Mystery Force Pushing Galaxies Away, 2002, Space & Science

Fairall, Anthony , Cosmology Revealed: Living Inside the Cosmic Egg, 2001, Praxis Publishing, Chichester, UK

Jones, L. , Stellar and Extragalactic Astronomy: How Stars Die, 2004, Physics Department, Gettysburg College, Gettysburg, P

Krauss, L. M., Energy and the Hubble Age", 2004, ApJ, 604, 481

No Author , Damages Caused by Atomic Bombs, 2004, mothra.rerf.or.jp/ENG/ Abomb/- History

Reddy, F., Tycho's Runaway Star, Astronomy, February 2005, pg 25.

Seeds, M. , Horizons, Exploring the Universe, 2004, Brooks/Cole-Thomson Learning

Villard, R. & Lloyd,, R., Astrophysics Challenged by Dark Energy Finding, 2001, Space.com

Weinberg, S. , Importance of Discovering the Nature of Dark Energy, Department of Physics, 2001,   University of Texas at Austin, http://supernova.lbl.gov/~evlinder/weinberg.pdf

Table 1   Summary of Measured Radial Star Velocities*

	Velocity Range 11000 velocities of 3500 stars (km sec^{-1})		
	Low	High	Isolated max
Southern Hemisphere	+0.2	+236	+555
(R.A. : 0 – 12 ^{h})	-1.0	-142	none
Northern Hemisphere	+0.4	+298	+346
(R.A. : 0 – 12 ^{h})	-0.4	-326	-414

*statistical analysis of data from Abt 1970

Table 2   Energy Released and Shock Strength of Various Violent Processes

Source of Explosion	Energy Released (joules)	Equivalent TNT (tons)	Shock Pressure (tons m^{-2})
1 ton TNT	4.20 E+09	1	--------
"Little Boy"	5.25 E+13	1.25 E+04	35
Atomic Bomb			
Supernova	1.00 E+46	2.38 E+36	?
Type II			

Note: A supernova Type II explosion will release more energy than 150,000 "Little Boy" atom bombs.

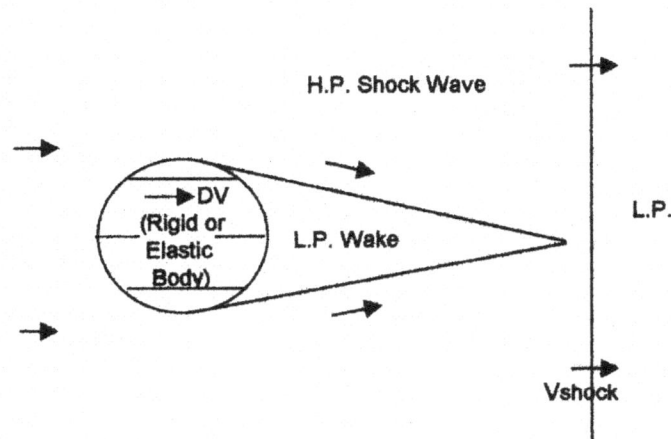

**Figure 1   High Pressure Shock Wave Creating Pressure
Differential on Star or Dark Body**

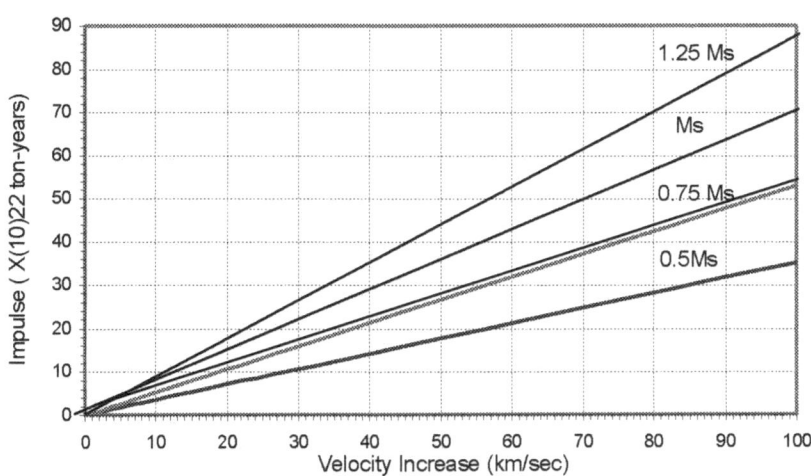

Figure 2   Impulse Required For Velocity Increase (0.5-1.25 Ms)

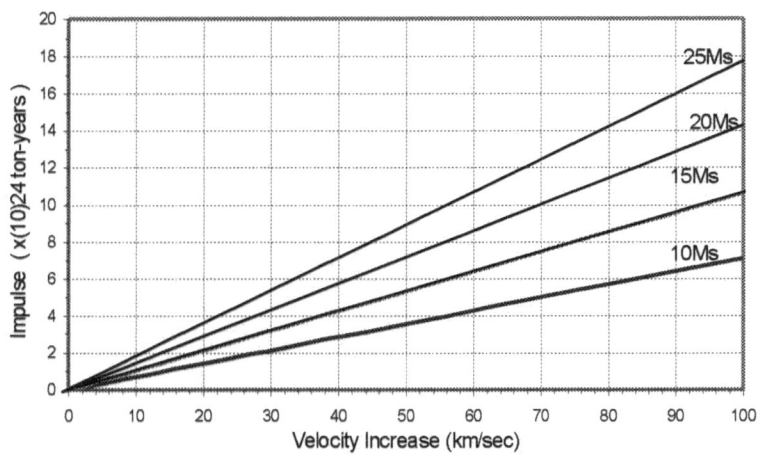

Figure 3   Impulse For Velocity Increase (10- 25Ms)

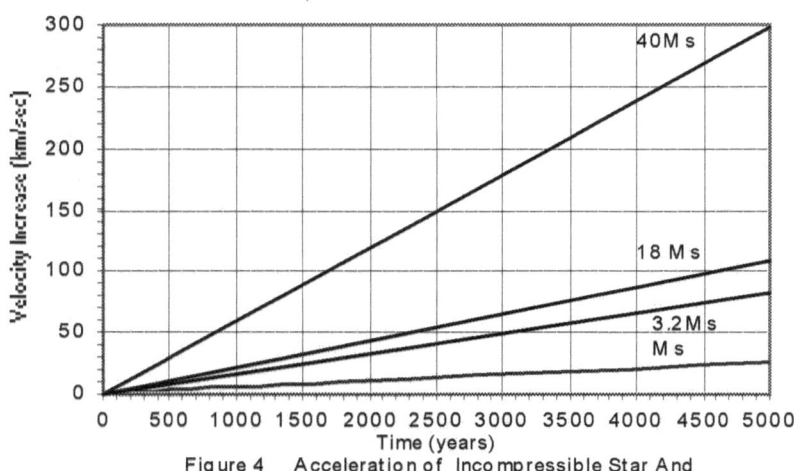

Figure 4   Acceleration of Incompressible Star And
Dark Bodies (P= 1 "Little Boy" atomic bomb))

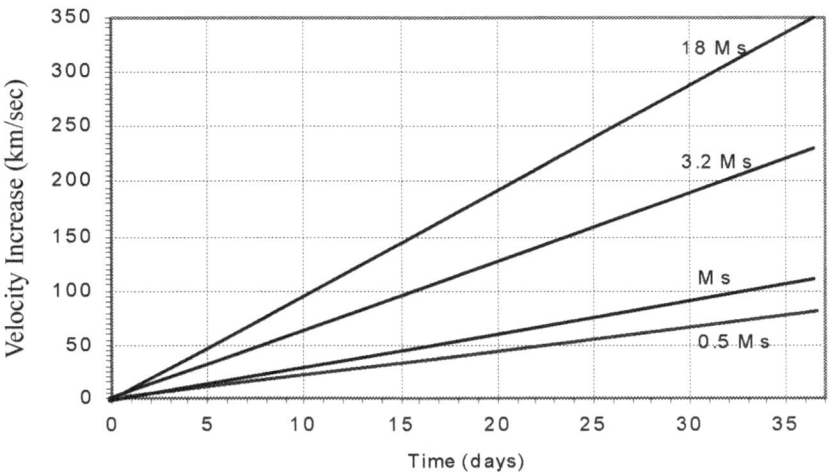

Figure 5   Acceleration Of Incompressible Stars And
Dark Bodies (P= 150,000 "Little Boy" atomic bombs)

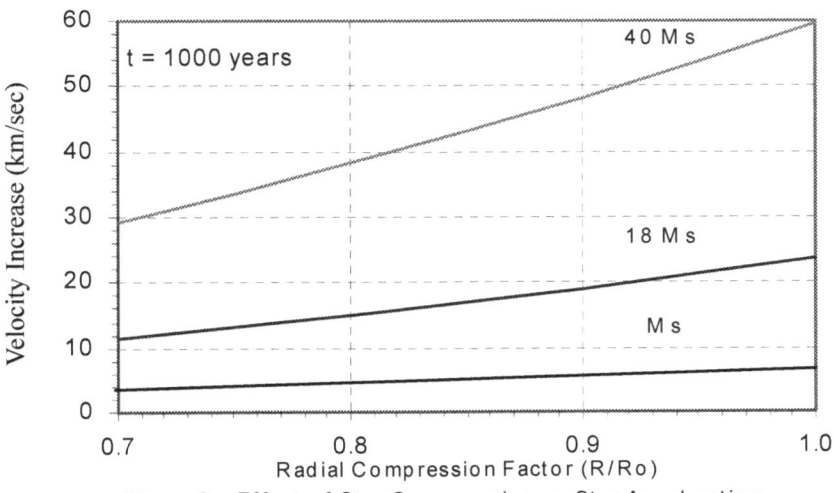

Figure 6   Effect of Star Compression on Star Acceleration
(P = 1 "Little Boy" atomic bomb)

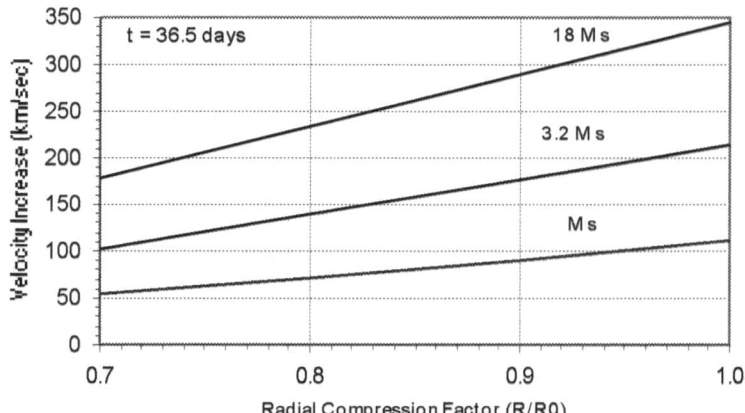

Figure 7 Effect of Star Compression on Star Acceleration
(p = 150,000 Little Boy" atomic bombs)

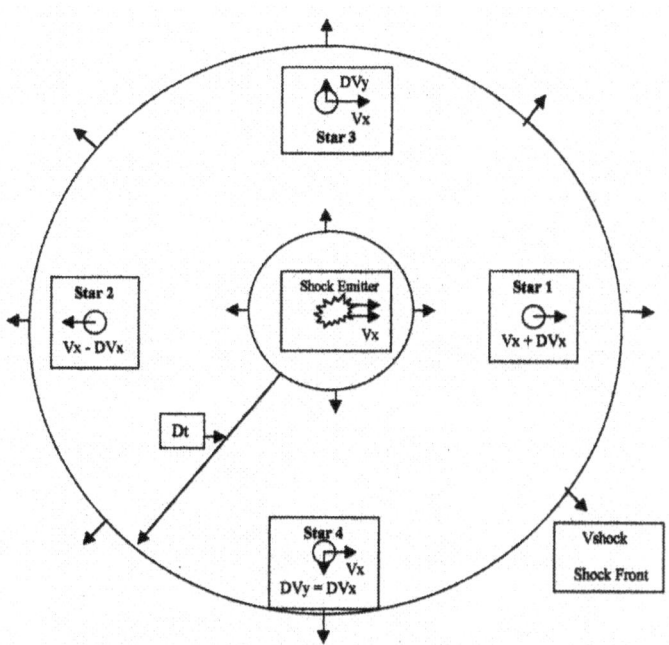

Figure 8 Spherical Shock Wave Moves Stars in Different Directions

Appendix

History

Of

Dark Energy Theory

STEADY-STATE CHARACTERISTICS

OF A

CROSS-FLOW ELECTRIC ARC

By

AUGUST ALBERT CEMKNER JR.

A thesis submitted to the
Faculty of the Graduate School of State
University of New York at Buffalo in partial
fulfillment of the requirements for the degree
of Master of Science

June 1969

March 18, 2004

Dear Dr. Cenkner,

SUBJECT: PROPOSED THEORY TO EXPLAIN DARK ENERGY

The editors want to thank you for letting them read your article.

Our editorial policy at Sky & Telescope is to report only scientific work which, by appearing in established research journals or professional meetings, has enjoyed some kind of peer review by the research community.

We suggest you send a thorough description of your theory to one of the primary research journals in astronomy or physics. Once your work has been accepted for publication in such a forum, we'd be happy to receive a copy of your refereed article.

Thanks again for taking the time to write.

Sincerely,

Lisa Johnston

49 Bay State Road
Cambridge, MA 02138-1200
USA

Telephone:
617-864-7360

Customer Service:
800-253-0245

Faxes:
617-864-6117 General
617-576-0336 Editorial
617-520-9518 Advertising
617-868-4481 Art & Design

Internet:
info@SkyandTelescope.com
SkyandTelescope.com

Editorial Assistant
Sky Publishing Corp.

Astronomy.

March 15, 2004

Dr. August Cenkner Jr.

▉▉▉▉▉▉▉▉▉▉▉▉

Dear Dr. Cenkner:

Thank you for submitting your article, "Proposed Theory to Explain Dark Energy." Unfortunately, we don't have a place for it in our packed editorial schedule.

As per our standard procedure we've enclosed the article you sent. Thanks again.

Sincerely,

Dave Eicher
Editor

21027 Crossroads Circle, P.O. Box 1612, Waukesha, WI 53187-1612
phone 262.796.8776, fax 262.798.6468
website: www.astronomy.com

August A. Cenkner Jr.

NEWSLETTER OF THE BUFFALO ASTRONOMICAL ASSOCIATION INC.

The Spectrum

A THEORY TO EXPLAIN DARK ENERGY – PART I

The Accelerating Universe

Dr. August Cenkner Jr.

Copyright March 7, 2004

Astronomers have observed that stars and galaxies, at the outer reaches of the universe, have actually accelerated and are therefore moving at higher, rather than the anticipated lower, velocities. Classical thought is that gravitational attraction, between celestial objects, would actually slow them down instead of speeding them up. They have attributed this unexpected behavior to some mysterious "dark energy", that apparently permeates the universe. The nature of this dark energy is presently unknown.

A dark energy theory is presented herein, that could explain the observed behavior of these astronomical bodies.

The basic theory states that multiple hypervelocity traveling gas-pressure/radiation-pressure/stellar-wind shock-waves are producing numerous force impulses when they impinge on an astronomical body. If the body is in front of the shock-emitter, these impulses would cause it to speed up. If it is behind the shock-emitter, it would slow down. The shock waves would be moving at a velocity that is higher than the emitter velocity, which would enable the shock to overtake bodies moving in front of the emitter. Bodies traveling behind the emitter would slow down because of retarding shock impulses.

This behavior can be predicted by applying Newton's Law of Motion. The unbalanced pressure on a star, that occurs upon shock impact, will produce a force impulse [force x time] that will increase the linear momentum [mass x velocity] of the astronomical body and hence increase its velocity and acceleration.

These traveling shock waves will also affect the velocity and path direction of nebula, (and possibly stimulate star formation) as these waves pass through the nebula. Nebula that are in front of the shock emitter would be accelerated while nebula that is behind it would be decelerated. Since there could be more than one shock-emitter in a galaxy or star cluster, these multiple shock waves would interact. The shocks could be reflected, attenuated or enhanced and could produce multiple or enhanced shock impulses.

The so-called dark energy is actually the energy contained in these high energy hypervelocity traveling shock waves. The thermal kinetic and internal energy of the star is converted into the kinetic and internal energy of the traveling shock wave. When the shock impinges on a star, or other body, it performs work on the body. This work increases the kinetic energy of the body.

Possible sources of these shock waves include exploding stars, pulsating stars, imploding stars, colliding stars, black holes as matter approaches the event horizon, stellar wind, etc. This shock wave energy would permeate the entire universe over time, since these shock-emitters would appear, eventually, wherever stars exist.

It has been reported that the core of the universe doesn't presently exhibit the presence of dark energy. Rather, it occurs in an expanding spherical shell around the core. This suggests that the stars are still aging in the core and they have to reach a certain level of maturity before they can become shock-emitters. Once they reach maturity in the dark energy zone, they are transformed into shock-emitters. This seems to suggest that star explosions and collapsing of mature stars may be the major emitters, with supernova and cataclysmic variables being the prime candidate as the most powerful of the relevant shock-emitters.

The theory suggests that, eventually, all shock-emitters will have been consumed. Gravity will again become the dominant force and drag may more significantly influence cosmic motion. The expansion of the universe will slow down, perhaps to the point that the universe eventually collapses upon itself.

Of course, some of the remnants of the shock-emitters may eventually coalesce into a new generation of stars. Some of these stars may then eventually transform into a new generation of shock-emitters. This would enlarge the dark energy zone and prolong expansion. *(continued on page 9)*

Table of Contents

A Theory to Explain Dark Energy – Part I ... 1

BAA Officials ... 2

BAA Web Site ... 2

Meeting Location/Time ... 2

Spectrum Deadline ... 2

Presidents Message ... 2

Upcoming Meetings ... 3

Convince Us, If You Can ... 3

College of Fellows Report ... 3

Funny, Funny, Funny ... 3

BAA Annals ... 3

Catastrophe, Catastrophe, Catastrophe ... 4

Spy and Tell ... 5

Astronomy Terms ... 5

Observatory Notes ... 7

BAA Policy ... 7

BAA Considers Joining Search for New Planets ... 8

Cartoons ... 8

A Dark-Energy Theory – Part II ... 9

Editors Corner ... 10

Subj: **Astronomical Journal - 204305**
Date: 7/23/04 1:31:36 PM Eastern Daylight Time
From: astroj@astro.washington.edu
To: jandgjr2@aol.com

Dear Dr. Cenkner:

The paper

"Universe Acceleration Predictions using New Dark-Energy Explanations" by
August A. Cenkner Jr.

has been reviewed by a knowledgeable referee, whose report is attached.

In view of the referee's report, I believe that the paper is not suitable

for publication in The Astronomical Journal.

Yours Sincerely,
Paul Hodge
Editor

Astronomical Journal
Dept. of Astronomy   Phone: 206 685 2150
University of Washington   FAX: 206 685 0403
Box 351580   astroj@astro.washington.edu
Seattle WA, 98195-1580   http://www.astro.washington.edu/astroj
----
I have read this paper and strongly recommend that the Editor return it to
the author with the indication that its subject matter is not appropriate
for the ASTRONOMICAL JOURNAL. It would be better submitted to a journal
concerned with fundamental physics and cosmology.

Sunday, July 25, 2004 America Online: jandgjr2

August A. Cenkner Jr.

Subj: **DARK2004 Conference Schedule**
Date: 9/20/04 3:41:43 PM Eastern Daylight Time
From: b-guster@physics.tamu.edu
To: b-guster@physics.tamu.edu

Dear Sir,

This email is to let you know that we are unable to accommodate your request to make a presentation at the upcoming DARK2004 Dark Matter in Astro and Particle Physics Conference, October 4-9, 2004. The conference is shaping up to be quite packed with invited speakers and so we are unable to add any others, however, we do appreciate your interest and hope that you can attend. I have attached a preliminary schedule for you.

Again, thank you for your interest.

Sincerely,

Beverly Guster

Beverly Guster, Program Coordinator
Mitchell Institute for Fundamental Physics
4242 TAMU
College Station, TX 77843-4242

Phone (979)845-7778
FAX (979)845-8674
b-guster@physics.tamu.edu

Tuesday, September 21, 2004 America Online: jandgjr2

UNIVERSE ACCELERATION PREDICTIONS USING NEW DARK-ENERGY EXPLANATION

August A. Cenkner Jr., Ph.D.

Buffalo State College, Elmwood Avenue

New Science Building

First Floor Auditorium, Room 213

Friday, September 10, 7:30 P.M.

In 1998, by studying the emission spectra from galaxies in the outer reaches of the universe, two independent teams of astronomers concluded that these galaxies have actually accelerated; they are moving at higher, rather than the anticipated lower, velocities (Ref 1-4). Classical thought is that gravitational attraction, between celestial objects, should actually slow them down instead of speeding them up. They have attributed this unexpected behavior to some mysterious repulsive force that apparently permeates the universe; it has been labeled dark-energy. The nature of this dark-energy is presently unknown. In addition, there are other currently unexplained phenomena like revolving star/dark body groups, galactic superclusters, voids, walls and nebula vortex. These phenomena will also be discussed and related to dark-energy. Finally, dark-energy will be exposed as a gas dynamic phenomenon and it will be shown how star acceleration occurs. This includes the results of laboratory simulations of dark-energy, that were performed at the University at Buffalo.

REFERENCES

(1) Villard, R. & Lloyd.R., "**Astrophysics Challenged by Dark Energy Finding**", Space.com, April 2001

(2) Chaikin, A., "**Dark Energy: Astronomers Still Clueless About Mystery Force Pushing Galaxies Away**", Space & Science, 2002.

(2) Weinberg, S., "**Importance of Discovering the Nature of Dark Energy**", Dept of Physics, University of Texas at Austin

(4) Krauss, L. M., "**Dark Energy and the Hubble Age**", Astrophysics Journal, April 1, 2004

August A. Cenkner Jr.

September 4, 2004

Professor Dale Taulbee
Department Chairman
Mechanical And Aerospace Engineering
315 Jarvis Hall, North Campus
University at Buffalo
The State University of New York
Buffalo, NY 14260

Dear Professor Taulbee:

I was inspired, by basic experimental research that I did in the School of Engineering at UB, more than 30 years ago, to formulate a theory to explain some here-to-fore unexplained astronomical phenomena. Also, all the gas dynamics education that I received at UB, including many courses that you taught me, gave me the insight to recognize the cause of these unexplained phenomena.

For years now, astronomers have been unable to come up with an explanation for these phenomena. I suspect that it's because many of them do not have extensive backgrounds in gas dynamics.

My gas dynamic theory is consistent with astronomical observational data and basic research that I performed with Dr. David Benenson, while working on my M.S. and Ph.D. degrees.

If I'm right – and of course I think I am – this is a major breakthrough in theoretical astrophysics.

I just thought that the School of Engineering might be interested in learning about this.

Yours truly

Dr. August Cenkner Jr.
Class of 66, 69, 73, 96

70

September 4, 2004

Editor
Lockport Union-Sun and Journal

Dear Madam/Sir:

I would like to make you aware of a potentially big local story.

The attached lecture will introduce a new theory, related to the evolution of the universe, to explain how some unexplained astronomical phenomena actually occurred. Astronomers have struggled, unsuccessfully for years, to explain these phenomena.

If I'm right – and of course I think I am – this is a major breakthrough in theoretical astrophysics and a big local story.

Yours Truly,

Dr. August Cenkner

Subj: JCAP/012A/0804
Date: 8/20/04 4:54:26 AM Eastern Daylight Time
From: jcap-eo@jcap.sissa.it
To: jandgjr2@aol.com

Dear Dr. Cenkner,

we regret to inform you that your paper "Universe Acceleration Predictions Using New Dark Energy Explanation" is not suitable for JCAP and has just been withdrawn from our journal.

Best regards,
JCAP Executive Office.

JCAP Executive Office - http://jcap.sissa.it
SISSA, Via Beirut 2-4, 34014 Trieste (Italy)
tel +39-040-3787571, fax +39-040-3787528

THE ASTROPHYSICAL JOURNAL

Logged in: August Cenkner (jandgjr2@aol.com)

AUTHORS: Check MS status

MS 61394

Title	Dark Energy Theory, with Laboratory Simulation, to Explain Universe Acceleration, Superclusters, Voids, Walls, Revolving Star/Dark Body Groups, Drag, and Wild Stars
Author(s)	August Cenkner, Jr.
Corresp. author	August Cenkner (jandgjr2@aol.com) [Change]
Proofs author	August Cenkner (jandgjr2@aol.com) [Change]
Editor	{}

Subjects   These subjects have **not** been verified by the journal office:

cosmology:theory-cosmology:miscellaneous-large scale structure of universe-shock waves

Versions

Version	Received	
1	29 September 2004	Access MS files	Upload additional files

Decision history

Version	Decision	Date
1	no decision yet	

Journal and Supplement: apj@as.arizona.edu   Tech support: apj-help@mss.uchicago.edu
Letters: apjletters@letters.as.utexas.edu

W3C WAI-AA WCAG 1.0   508 BOBBY APPROVED

About the Author

Dr. August A. Cenkner Jr. has an interdisciplinary background that includes the areas of gas-dynamics/physics/astronomy/computers. He earned a B.A. degree in computer science, a B.S. degree in aerospace engineering, and M.S. and Ph.D. degrees in engineering science. While concentrating his graduate work in the area of gas-dynamics, he also took a number of advanced physics courses – i.e. Supersonic Flow (1&2), Hypersonic Flow (1&2), Non-Steady Gas Dynamics, Rarefied Gas Dynamics, Molecular Flow, Radiation Gas Dynamics, Radiation Heat Transfer, Aerothermochemistry, Magnetohydrodynamics, Plasma Physics, Electromagnetic Theory, Electrodynamics, Advanced Mechanics, Quantum Mechanics, Direct Energy Conversion, Diagnostic Techniques, Space Science and N-C – Spectroscopy and Optics.

For his M.S. thesis and Ph.D. dissertation, he respectively conducted original experimental spectroscopic research on cross-flow and coaxial plasma wind tunnels. His eleven-year part time tenure as a lecturer, in the School of Engineering, included teaching dynamics, fluid mechanics, solid mechanics, and thermodynamics. During more than thirty years in R&D, he was involved in research on gas dynamic lasers, shock tubes, supersonic/subsonic nozzles and diffusers, supersonic/subsonic mixing, compressors and various types of wind tunnels. He has authored or co-authored twenty-one publications.

Dr. Cenkner is a long time amateur astronomer; he has studied astronomy extensively.

www.ingramcontent.com/pod-product-compliance
Lightning Source LLC
Chambersburg PA
CBHW022114170526
45157CB00004B/1636